THE
TAILORED
BRAIN

FROM KETAMINE,
TO KETO, TO COMPANIONSHIP,
A USER'S GUIDE TO
FEELING BETTER AND
THINKING SMARTER

THE
TAILORED
BRAIN

EMILY WILLINGHAM

BASIC BOOKS

New York

Basic Books
Hachette Book Group
1290 Avenue of the Americas, New York, NY 10104
www.basicbooks.com

Printed in the United States of America

First US Edition: December 2021

Published by Basic Books, an imprint of Perseus Books, LLC, a subsidiary of Hachette Book Group, Inc. The Basic Books name and logo is a trademark of the Hachette Book Group.

The Hachette Speakers Bureau provides a wide range of authors for speaking events. To find out more, go to www.hachettespeakersbureau.com or call (866) 376-6591.

The publisher is not responsible for websites (or their content) that are not owned by the publisher.

Print book interior design by Amy Quinn

Library of Congress Cataloging-in-Publication Data
Names: Willingham, Emily Jane, 1968– author.
Title: The tailored brain : from ketamine, to keto, to companionship, a user's guide to feeling better and thinking smarter / Emily Willingham.
Description: First U.S. edition. | New York, NY : Basic Books, 2021. | Includes bibliographical references and index.
Identifiers: LCCN 2021021510 | ISBN 9781541647022 (hardcover) | ISBN 9781541647015 (ebook)
Subjects: LCSH: Cognition. | Brain. | Thought and thinking. | Happiness.
Classification: LCC BF311 .W5883 2021 | DDC 153—dc23
LC record available at https://lccn.loc.gov/2021021510

ISBNs: 9781541647022 (hardcover); 9781541647015 (ebook)

LSC-C

Printing 1, 2021

To all the beloveds in my life,
and especially to my mother,
who from my earliest days opened
door after door to opportunity for me.

CONTENTS

INTRODUCTION

I AM GLAD you are reading this book because it means that your brain and my brain are interacting. Using your brain, you are encountering things I had in my brain, and my hope is that at the end, you will feel that the interaction was useful. The preceding sentence encapsulates the main theme of this book: when human brains interact with each other, they offer benefit beyond the individual. What distinguishes this book from many others about "improving the brain" is that the improvements aren't for your benefit alone. As a social species in easy connection with each other, we humans have the best tools available to tailor all our cognitions: global, social, stress, attention, mood, and creative. In doing so, using one of the oldest tools we have available—one another—we create benefit for all of us.

When people talk about improving their brains, especially self-improvement, they are usually thinking about chemical tweaks, superfoods, drugs, electricity, magnets, supplements, herbs, playing chess, listening to Mozart (it is always Mozart), or nature bathing among any trees that haven't burned down yet. But the things that genuinely work to improve our cognition, creativity, memory, attention, and mood aren't usually pills or even fun and challenging games. Sure, keto diets, ketamine, and brain-stimulation techniques can have effects, especially on specific ways we use our brains. But what they do often doesn't compare in magnitude, persistence, or breadth to the brain-wide impact of the human factor. Yes, one of the interventions that works best tends to be a little closer to home, sometimes literally, because the secret ingredient is being a human around other humans.

Lots of books and people flogging supplements and brain "rewiring" promise that they'll give you the brain you want. But most of these interventions may well target the wrong outcome. What, really, is the benefit of a "brain booster" that promises to bump your IQ three points versus a strategy that might help you become a kinder, more understanding, more empathetic person? Even though an essential feature of our species is being social, empathy and social-cognition skills don't ever seem to be at the top of the "to tailor" list for our brains. When is the last time a supplement manufacturer promised you that their product would make you more empathetic or kinder? What programs promising to "rewire your brain" target such an aim, unless it's so you can manipulate others to your purposes? Certainly, such programs don't talk about how *you* can use enhanced tools like empathy to help *someone else* have a better brain. This book, however, does.

Here, I use peer-reviewed evidence to characterize mostly self-driven and next-generation approaches that might be truly useful in customizing a brain you'd be pleased to have. The options cover ingestibles, electrostimulation, diet, lifestyle practices, and taking the measure of your social cognition. If you're intractably committed to the aim of being "smarter," shaping your social-cognitive capacities can offer better overall cognition as an accoutrement. It turns out that the work of understanding and interacting with other people can be our best brain-tailoring tool.

We start by mapping out our Planet Brain and the infrastructure that keeps it functioning. The planet metaphor I use here is intended to create a spatial memory for you. With that mental map, you'll find it easier to navigate the complex, forgettable, and often redundant names we use for brain anatomy and brain function, along with the many layers of brain architecture, from lobes to nodes to neurons. By the end of Chapter 1, your map will be a handy internal reference when you run into terms like "insula" and "parietal" and "salience network." In addition to explaining why it's not good for the brain to smell like goat urine, this chapter tracks the particular arrangement, contours, and connections of Planet Brain, providing a basis for understanding how the interventions detailed in this book might work—or not work, as the case may be. The chapter ends on the book's primary theme: your Planet Brain is part of a system. It's not a biosphere that stops just behind your skull, but a natural component of a larger interacting collective of other brains, all in constant states of change. Your best bet for function in that system is to build internal fluidity for external flexibility, not by focusing rigidly on yourself and your own brain but by developing and maintaining healthy connectivity with others.

The interventions we look at in this book can rely on complicated technology, as can the studies that evaluate them, and these therapies act on our most complex organ at different levels. The first chapter is intended to orient you to the physical environment of the brain. The second chapter orients you to what these interventions do and how they do it. Understanding the relevance of research into these approaches means needing to understand the studies themselves—and the claims that are made based on those studies. So Chapter 2 also includes some clear, accessible information about gauging evidence and how to tell when that evidence is solid or flimsy. It explains the tools in the tailoring kit before we forge ahead to Chapters 3 and 4, which take on the enormous question of global cognition.

Most people interested in "brain training" for their individual betterment would probably consider these chapters to be key for them. Indeed, plenty of books, marketing materials for brain-training outfits, and supplements on shelves target "enhancing cognition," which remains the most sought-after prize in the brain-improvement industry. In these two chapters, you'll learn about the damaged and dark history of how we've tried to measure the elusive trait of improved cognition and the motivations lurking beneath those efforts. Then you'll find an assessment of the evidence for various claims around computer- and video-based brain training, transcranial stimulation, meditation, and more. I hope the content of these chapters surprises you, perhaps about yourself and certainly about what we think we're improving and why. You'll definitely find a few interesting, short cognitive challenges to take for a spin (one of them with a 0 percent solve rate in studies), which are always fun. This pair of chapters closes with an examination of the interventions that seem to yield the most bang for the buck when it comes to cognitive benefit. Perhaps at this point, you won't be surprised to learn that the big success story doesn't involve only a single individual brain but the influence of many brains working together to solve a problem.

That takes us to Chapter 5 and a look at how to improve our social cognition. This essential chapter discusses what is crucial about this ability and why it's important for helping our own brains and those of others around us. We will examine empathy and its biological underpinnings, along with what can impede our abilities (perhaps not surprisingly, being socially isolated does not help) and what we can do to refresh and reshape them, from up close or far away. Given the experiences of the twenty-first century to date and the social fractures that have left our world feeling intensely unstable, you'll find good news here: You can do lots of different things to improve human connection and understanding. In turn, your brain and the brains of others will benefit, including in that one way most people think they need—overall cognition. But the first step is identifying where the seams are and how to reinforce them. This chapter is the core of the book, with information that I truly believe could make a difference beyond the scale of one person. It begins when we recognize the power of storytelling.

Speaking of stories and the twenty-first century, plenty has happened to put most of us on edge, with stress and anxiety fogging up our capacities. In the next pair of chapters—on stress and anxiety and on attention and memory—we look at the associations of brain structure and function with these states, including diagnosable conditions. Then we dig into the accessible interventions that show promise or have confirmed benefit, especially highlighting those that people can try on their own. I don't spend a lot of time here on prescription drugs and standard behavioral therapies, which have been amply covered in dozens of books. In these chapters, you will instead see that the concept of "tailoring" the brain with accessible tools really stands out. Our feelings of stress and anxiety, memory changes, and attentional struggles are extremely subjective, so these two chapters are highly personal. Yet as we add instruments to our tailoring kits to enhance what I refer to as our stress and attention cognitions, you will find that the most successful of them involves the human factor that surrounds us.

The next chapter, on mood, is the most serious. Compared with the rest of the book, I spend far more time in Chapter 8 digging into information about interventions that cannot be tried at home. Most of these therapies require professional supervision because depression in particular is associated with some of the direst outcomes and is highly resistant to many nonclinical approaches. People with depression must clear several hurdles before they can get to the effective supervised interventions I focus on in this chapter. My hope is that evidence of effectiveness might motivate earlier access for this population. The chapter also features an examination of how psychedelics are faring in studies of their effects on mood because these drugs are exciting a high level of interest in this area.

After this trio of sobering chapters, we get into something a little lighter and particularly human: creativity. Of course, psychedelics get a treatment here, too, as do various other methods that are claimed to enhance the multicolored blossoming of the creative human brain. Although acid can take you on quite the journey, it is usually a solo adventure. There's nothing wrong at all with exploring the farthest reaches of your individual ingenuity, and there are plenty of ways to do that. But what's the upshot of those discoveries? As this chapter emphasizes, for a social species like ours, the mediators and beneficiaries of our creativity tend to be the same thing: other people. When we create something and share it, that creation connects us to others, and the shared experience and response highlight our commonalities. Empathy emerges powerfully around our creations.

What better to follow a chapter on creativity than a peek at the wild, wild future, which is all about human ingenuity? We've got pigs with a thousand tiny robot-implanted electrodes to tell us when the animal is smelling something, and a burgeoning field of research into connecting with others in social brain nets. Whether these trends mean we're facing down a utopian collective or, as Elon Musk put it, a *Dark Mirror* episode remains unclear.

But I don't think there will be a pill, red or blue or purple, to turbocharge our brains. The way to being smarter, as a population or individually, isn't going to arise from exploiting some brain supercomputing power. Despite movies such as *Limitless*, which features a chemical that can make you a wildly brilliant Bradley Cooper, that scenario's not likely either. We can be smarter, as individuals or collectively, if we embrace what makes us human: our capacity for sharing information in the form of stories that drive memory, attention, empathy, imitation, problem-solving, and emotion recognition, and for building on the fruits of many minds. If we look away from that and focus only on ourselves, we risk becoming inhuman.

In the end, this book about tailoring the brain should really be considered a book about tailoring the *brains*. Even interventions that shore up how our personal brain operates also facilitate how we interact with others. In reading this book, in fact, you and I are interacting. If you gain benefit from it—whether from insight, interventions, or both—in shaping the feeling of a mind at peace, then this book itself has produced a positive effect from the interaction of two brains. That's a pleasure to contemplate.

In his Pulitzer-winning book *So Human an Animal: How We Are Shaped by Surroundings and Events*, René Dubos wrote that the essential nature of humans hasn't changed since the Stone Age. I think that Dubos was right. Our fundamental makeup has not changed since that period, when our social savvy likely was crucial to our survival, as it still is. But ours is a multifaceted nature. There's no one weird trick that will work on every person or even on the same person across a lifetime. We contain multitudes, and we are shaped to be a part of multitudes. My hope is that the approaches you find here will lead you to a brain at peace with itself and the other brains around it. We all could use a little more peace.

CHAPTER 1

MEET YOUR BRAIN: THE PLANET

THE FIRST KNOWN recorded mention of the brain is on a papyrus that dates to about 1600 BCE. It's actually a copy of something first written around 3000 to 2500 BCE in Egypt. The document is dubbed the "Edwin Smith Papyrus" because Edwin Smith, an antiquities dealer and Egyptologist, bought it off someone in 1861. Made in an era without book binding, it stretches to 4.7 meters. The person who transcribed it from the lost-to-time original was perhaps not the most meticulous at their work. But the series of glyphs in hieratic clearly details the medical woes of several people who sustained head injuries, and the treatment, if any, they received. In these case studies lies a reference to the human brain. The glyph looks like Figure 1.1.

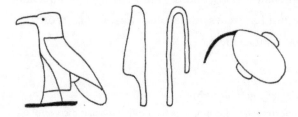

Figure 1.1. This is "brain" in hieratic.

The text also includes a list of features for the physician to examine in a patient with suspected brain injury, in order and with key clinical signs, along with ancillary characteristics to check. The list is thorough and includes the queries "Does placement of the fingers in the mouth of the wound lead to great disturbance?"* and "Odor from the calvaria [skull] similar to the urine of goats?," which would signify infection. There is mention that the brain tissue might pulse, like a child's fontanelle.† Neurological symptoms, such as an inability to speak, were on the list, too, showing some understanding that brain injury and these problems were linked.

This ancient diagnostic checklist offered a guideline for the physician to decide whether they could treat the patient or not. If the latter, nothing would be done. For treatable damage, the therapies could sound worse than the wound. For a lesion perforating the skull, fresh meat (origin not specified) was to be applied for twenty-four hours, followed by placement of the patient on the ground "until the time of his suffering passes." If the wound penetrated to the skull, the prescription was daily "binding with grease and honey and lint" until recovery. For the unfortunate man (all the case studies were men) who sustained a cracked skull, torn membranes around the brain, and a bruise to the brain itself, the diagnostic process involved "palpation of the wound." A frontal penetrating stab wound accompanied by signs of infection or tetanus was deemed (accurately) as a "medical condition you will not be able to treat." The action was simply, "No recommendation."

A BRAIN SHOULD NOT SMELL LIKE GOAT URINE

Modern clinicians have declared themselves struck by the "astonishing observational skills" these ancient physicians demonstrated. They understood the gross anatomy of the brain, the association of heartbeat and pulse, the meaning of injuries in different regions, and the ominous signs of infection. Their diagnosis and classification were not, in some cases, unlike those that would be made for similar injuries today.[1]

---

* Just reading the question gives me a great disturbance.

† The "soft spot" of the skull not covered by bone. It persists through the first months of life, before the skull bones suture together.

The document also gives us the earliest known printed comparison regarding the brain. If the clinician were to find "ripples like those in copper in the melting process and something within that throbs and flutters like a child's fontanelle," that was a bad sign. Verdict: "This is a medical condition you will not be able to treat."

Things have changed, among them the ways we liken the brain to visual references. Most of us don't have a lot of experience observing ripples in melting copper, making that comparison somewhat mysterious to us.* The perception Egyptians held of the brain may have influenced the choice of simile: they viewed the brain mostly as the source of mucus, something to be removed through the nose at death. Other, more important organs were carefully preserved in separate vessels.

Over time, the brain has been metaphorized as a lot of things. These analogies have changed as we have gained understanding of the brain and the rest of the natural world. People once viewed the brain as a blowhole for releasing heat to cool the heart, the true seat of consciousness, per Aristotle, and as a repository of the soul—specifically in the fluid-filled gaps called ventricles. We've dispensed with the idea that it's an organ made entirely of marrow or even of sperm (!), which would travel thence to the testes by way of the spinal cord.† We've survived a theory of humors as the basis for brain diseases, with "phlegm" being responsible for epilepsy, by preventing necessary "vapors" from entering and refreshing the blood, leading to seizures. We've even come back around to the notion of the release of vapors, this time through the sutures holding together the bones of the skull.[2] One thing that seems unchanging is that many of us can sometimes feel an intense sensation of pressure in our heads that needs some release.

And there are, of course, all the mechanized references: gears, clocks, machine, wiring, rewiring. The brain has been a tiny mechanical puppet, a machine made of springs, or a telephone switchboard, unplugging and

---

* Curious about this visual, I watched videos of people melting copper, which is a surprisingly robust corner of YouTube, and it seems to me that it might be a reference to feeling a subdural hematoma, or blood-filled clot, on the raised ridges, or gyri (singular "gyrus"), of the brain surface.

† No word on what this meant for people who didn't produce sperm.

replugging as it moves from thought to thought. It is a calculator, a computer with storage capacity, a jumble of wires that can connect and reconnect, sometimes by mere exposure to a bad television show. It has been thought to gate logic at yes/no decision points and then to continue on its path, mechanically and algorithmically. We humans loved that one because to us, logic and its companion, reason, are the sine qua non of humanness, features that our brains manifest with exemplary superiority.* The brain is somehow even "like a scientist," generating hypotheses or predictions and then testing them against incoming signals[3]—a tautological comparison, of course, and also made for the wrong reasons. Scientists are humans, with all the faults that pertain thereto.

You'll notice a trend. We keeping taking these living, ever-changing, natural, emergent properties and reducing them to a single aspect of complexity. None of our comparisons encompass the whole caboodle. It's interesting how we instinctively step away from looking at our brains as something alive and vulnerable, as much a part of nature as grass. Instead, we frame them as separate and distinct from the natural world, mechanical and wired things. We don't often substitute mechanical metaphors for trees or penguins.† Why do we do that with the alive, squishy, grey-pink mass in our heads? Because it can do math? What we leave out is that to do math, we engage with the outside world that we're quantifying, trying to capture its contours in this reductive way. But in our expression and behavior, we never, ever stop at pure numbers.

Each of these metaphors omits the cause-effect-cause chain of interactions between us and the world. That world includes the other human brains in our orbit, all natural entities contributing to the natural environment. Most comparisons reference what we put into our brains and what happens *internally* after that. But what about our output, the reactions we and others have to those inputs? That is, after all, what builds a community of brains, connecting us with the rest of nature.

---

* They do not.

† Although we are prone to describing trees as "lungs of the earth" and to comparing penguin plumage to a tuxedo.

In 2002, Paul G. Allen, originally famous as the cofounder of Microsoft, funded a project that came to be known as the Allen Brain Atlas.* The idea was to develop a three-dimensional atlas of brain anatomy and gene expression. Tie in function—what the brain does, when, and where—and the layers can produce a map of the three facets combined. The project points to a metaphor for understanding, remembering, and talking about the brain that does not erase it as a construct of nature, intended to interact with the world outside its casement and with other brains in that world. The mapping project takes something that happens in time (the brain's use of specific genes under specific conditions) and adds space (*where* the genes are being used). It's a three-dimensional genetic information system† for mapping Planet Brain.

Like planet Earth, the living, mappable Planet Brain comes complete with continents, countries, cities, citizens, nutrient cycles, and energy needs in constant flux, all facilitated and connected by a rapid transit system of communication along a flexible but vulnerable infrastructure.

PLANET BRAIN

Let's take a journey, starting from a starship-level view, of the planet that is our brain. It's like no other organ you have. Most of our organs are built from distinct layers and compartments with specialized cell types that have well-delineated roles: on-off binary processes, homeostatic constraints, limited redundancy,‡ relatively limited complexity.§ These organs tend to operate within narrow boundaries, with little leeway for error or excursions beyond normal without significant risk of permanent damage or death. We know the outcomes of damage to tissues and can predict with fair certainty what that damage implies. If your lungs develop scarring, your lung capacity will probably be diminished, and you'll become breathless sooner with exertion.

---

* Housed online at http://human.brain-map.org/, along with a half dozen other atlases for different stages of development and species.

† When we do this for Earth, the GIS stands for "geographic information system."

‡ Even when, as with the vessels that serve the heart, a little redundancy would be lifesaving.

§ I know that organ-specific enthusiasts will quibble—"but the kidney!" and so on—but it's still true.

But our brains are different in almost every way. You might think that after millennia of attention to this organ, we'd know and understand it in intimate detail, but we do not. We are nowhere close to completely comprehending most aspects of the human brain. We haven't even fully mastered its anatomy, which we've been able to study for a long time. Get into cell types and what they do, what they make, when they make it, and why, and it still feels like we're very early in the learning process. Try to tie anatomy or cellular understanding to function and how it all comes together so that we can do what we do, and you will find little in the way of broad consensus about most topics.

Only fairly recently have we gained enough information to debunk widespread myths about the brain, such as "we use only 10 percent." As with our ancestors and their ideas about vapors and humors and sperm brains, we of more modern times have made some major blunders of our own. Have you heard that the brain has a hundred billion neurons, or that some people are "left-brained" and some are "right-brained," or that our brain is special because it has developed beyond the "lizard brain" of "lower species"? These are ideas that lived and breathed in this century. And they are dead wrong. As far as we currently know.

SIDEBAR: Now is as good a time as any to deliver this caveat: obviously, our understanding of the brain is a work in progress, undertaken by the very organ we are trying to understand. Technological advances and the advent of standardized research designs (sort of) add a layer of certainty to what we're learning today. But that layer, like the "left-brain/right-brain" idea, can crack too. When I talk in this book about research findings, keep in mind that the vast majority are just inches gained (or lost, depending) on an ongoing journey, and they still fall short of the ever-elusive destination: a ground truth about the brain.

PLANET BRAIN: THE CONTINENTS

Let's explore what we do know.

Put your hand on your forehead, as though you were checking for a fever. Just behind your hand (well, behind the skull behind your hand) are your two frontal lobes, one on the left, the other on the right. These are your gatekeepers, tempering emotions and judgment, solving problems, and, if you're lucky, operating on your behalf as mature adults.

At the leading edge of each lobe is a region called the prefrontal cortex. It gets its name from its location: it lies at the front of the frontal lobe proper and is part of the cortex, the outer portion of the brain. In its adult form, it gates behaviors that we associate with maturity. Personal features that seem immutable about an individual can change when the prefrontal cortex is damaged. An example is the famous and famously misunderstood case of Phineas Gage, who ended up with a 13.25-pound iron spike—measuring 1.25 inches thick and more than a yard long—impaled through his left frontal lobe.

Gage has been described as shifting from being a pretty nice fellow to acting like a coarse brute, a somewhat dramatized version of his story. The reality was that after his harrowing experience,* he regained enough function to hold down a job in Chile before his wounded brain sustained intractable seizures, from which he died in 1860 at age thirty-six. The real interest of the story is the fact that he survived the injury, its horrific aftermath, and a subsequent infection, and lived for another twelve years, a testament to the resilience of the brain beyond that of other organs.

Reports vary somewhat, but Gage's personality and impulse control seem to have changed in retrograde. In people whom you might view as being in a state of "arrested adolescence," the prefrontal cortex is where the arrest happens. The two decades or so that are required for this part of the brain to mature—it is the last of our lobes to do so—mean that people around age eighteen to twenty-two are especially primed to act before fully processing incoming information. Knowingly or not, military drafts through time have leveraged this situation to ensure recruits in the prime of physical strength and at peak insouciance about consequences. Yet as dire as the damage was that Gage sustained, his physician essentially wrote that if you're going to

---

* The spike entered his head from below his jaw, passed behind his eye, and exited through his left frontal lobe. He could still talk after it happened. After surviving what sounds like a horrible infection arising twelve days later that involved brain tissue pushing out of his wound, by day twenty-four he was walking. He worked for a few years before the seizures developed.

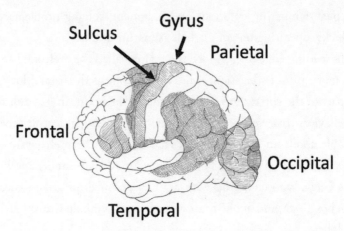

Figure 1.2. The lobes of the cerebral cortex, with an example gyrus (in this case, the somatosensory cortex, shaded) and sulcus (central sulcus).

be injured by an iron spike jamming through your brain, the front is better than the back, where brain functions crucial for life are seated.

Let's edge a little toward that back part of the brain. At the rear border of the frontal lobe, perhaps where you might wear headphones during a Zoom call, is an area of the cortex that coordinates your movements. It's a raised region, called a gyrus, and it spans from ear to ear, like an undulating arch. This motor (movement) cortex contains neurons that send out messages about how to respond to your environment—and so you move. Across the span of this cortical strip, neurons are arranged systematically to process information from specific parts of your body, giving complete coverage to the left side and the right side. In this way, information related to your toes is always processed in the same regions on the left and right sides of the span (in this case, toward the top of your head).

Inching toward the back of your head just a bit, we encounter a shallow canyon, or sulcus* (Figure 1.2). This central sulcus separates the frontal lobes from the parietal lobes. Once we cross it, perhaps swimming through the cerebrospinal fluid that fills it, we have entered the territory of the parietal lobes. At the leading edge of that land is another arch of brain, or gyrus, called the somatosensory cortex. This region is responsible for processing

---

* Plural "sulci."

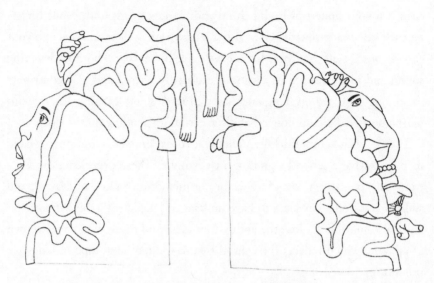

Figure 1.3. The order of outgoing/motor signals and incoming/sensory signals by target body region. The left side shows the distribution in the motor cortex, and the right side shows the distribution in the somatosensory cortex. In a real brain, the motor and somatosensory cortexes span the entire brain, each covering the left and right sides of the body. Redrawn after Wilder Penfield.

sensory information, and it confers with the motor cortex about how to respond.

Sensibly, inputs enter this brainspan at specific regions for specific areas of the body, in alignment with the same pattern on the motor cortex. As you may have guessed, this alignment means that where toe inputs map in the motor cortex (near the top of the head), they also map to an adjacent region in the sensory cortex. That way, the two brainspans can confer directly, so that you can wiggle your toes and feel them after you read this. Figure 1.3 shows a common depiction of the order of these regions on each of the bands. The more neurons devoted to the region, the larger the body part as shown in proportion to the rest.

The frontal and parietal lobes lie above another set of lobes, separated by another canyon, the lateral sulcus—or, more beautifully, the Sylvian fissure.* With your two hands, rub your temples, as though you had a stress headache;

---

* A fissure is a deeper, wider version of a sulcus.

then run your fingertips back behind your earlobes. Just under your fingers on each side is a temporal (as in, at the temple) lobe. Their adjacency to your ears is your tip-off that these lobes have a role in hearing—both in detecting sound and in processing its meaning. Relatedly, these lobes also contribute to, among other things, language-based processes and learning and remembering both verbal and nonverbal information (depending on the side).

People with temporal lobe epilepsy can experience symptoms related to hyperactive electrical signaling in this lobe. These experiences include (transient) confusion, loss of a memory or memories, a sense of déjà vu, and sudden emotions that seem to have no trigger. People with this kind of epilepsy also may briefly lose the ability to understand or use language, known as aphasia, which reflects the central role of the temporal lobe in language processing.

Continuing our journey to the opposite pole from where we started at the front, put a hand on the back of your head. You're covering your occipital lobes. These small but mighty lobes are the seat of vision; they are involved in both detecting visually and processing what you've detected. That's why getting bashed on the back of the head can affect vision. Seizures, or electrical hyperactivity, confined to this lobe can produce visual hallucinations. To reach the occipital lobes from the parietal, we had to cross yet another canyon, the creatively named parietal-occipito sulcus.

The temporal and occipital lobes sit atop another crucial brain structure, one that is often overlooked when we talk about thinking, creativity, and executive function (our ability to make choices to achieve our aims). That structure, whose layers make it look like a topographic map, is the tough cerebellum, a beautiful and beautifully organized region of our brains. It's not giant and flashy, like the cortex, but it's lovely and richly complex, the New Zealand of Planet Brain, minus the hobbits. Despite its location underneath the parietal and temporal lobes, the cerebellum—historically viewed as having only important but unconscious duties associated with balance and such—has strong connections with the overlying cortex, including with the prefrontal cortex, the seat of mature humanity.

As you probably are aware, like any decent planet, the human brain has hemispheres, a left and a right, connected by a tough bridge of tissue called the corpus callosum ("tough body"). Also tucked away deep* within our brains is the limbic† system, a series of stacked, paired structures that gate and relay information up and down Planet Brain. These structures do too many things to detail here, but let's highlight a couple that you'll see again and again.

The hippocampus‡ has a central role in the formation of memories, connecting them to our senses, so that the sight, smell, taste, or feel of something calls up a related, stored memory from elsewhere in the cerebral cortex. The hippocampus can be where a seizure begins in temporal lobe epilepsy: some of its excitable neurons that connect to the temporal lobe become uninhibited and fire at will, creating a temporal storm. With the role of the hippocampus in memory, this behavior may explain the déjà vu that can accompany such seizures.

The amygdala, which lies next to the hippocampus,§ paints those memories with emotion. Some have interpreted the movie *Inside Out* as depicting in part the limbic system, including the amygdala and the limbic-associated insula.[4] Each of the five emotions in the eleven-year-old protagonist's brain could tinge a memory ball with the color of the emotion it represented.¶ Memories that we process with strong emotion tend to take on special clarity, which is one reason I try to make my students laugh when I am teaching them. If the emotion is strongly negative, it can be associated with posttraumatic stress disorder. The amygdala and the hippocampus work in tandem, and both have two-way connections with the cortex. In fact, the amygdala is also involved in temporal lobe epilepsy, with a role in the extreme feeling of joy or fear that can rise with the seizure event.

Other parts of this deep area of the brain work with hormones and other molecules in feedback loops to govern everything from digestion to growth to making gametes.

---

* In anatomy, if you are "deep," you are far from the surface, which is "superficial." Anatomy often reflects real life this way.

† Latin for "related to a border."

‡ Greek for "seahorse" because of its shape.

§ Remember that all these structures occur in pairs.

¶ "Sadness," for example, could tint a recollected happy memory blue, a sign of the bittersweetness of nostalgia.

In our brain-as-planet metaphor, each part of that planet—the lobes, the cerebellum, the myriad and distinct elements of the limbic system—is a continent, a region of terra firma separated from other regions by canyons and gaps filled with fluid, much like the continents on planet Earth. And within each of the lobes lie smaller regions that are like countries, each with a distinct boundary and culture, at least from a distance, often communicating with each other across those distances. You've already ventured into a few of these territories, having crossed the motor, prefrontal, and somatosensory cortexes.

PLANET BRAIN: THE COUNTRIES

Like the real geography of planet Earth, the borders delineating countries in the mammalian brain have shifted around through time* (Figure 1.4). Chimpanzees and humans (and elephants and dolphins, among others not shown) have a lot of those canyons (sulci) and ridges (gyri), whereas in many other mammals, the brain's surface is relatively smooth. As time and natural

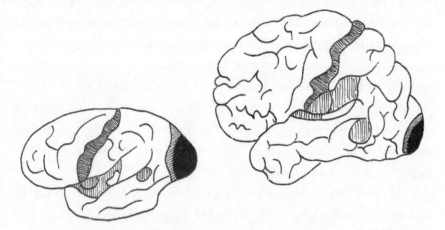

Figure 1.4. The distribution of areas of the cortex for different functions, and the relative sizes of these areas to the rest of the brain, as it differs between chimpanzees (left) and humans (right). Solid black in the back (occipital lobe) indicates the visual area, and the hatched region just before it is a second visual area. The hatched spherical area is the primary auditory area, and the hatched gyrus going over the brain is the somatosensory cortex. After Buckner and Krienen 2013.

---

* And they also shift around within our own brains as we age and connections change.

selection eventually gave the world the great apes, including us, the borders of regions such as the visual areas (in the occipital lobe) and the somatosensory area or cortex (where the parietal and frontal lobes meet) tightened up quite a bit relative to the size of the brain itself.

So our Planet Brain differs from the Planet Brains of other mammalian species quite a bit in its contours, geographical borders, and proportional allocation of each country or region. Nevertheless, you can see that the broad pattern persists. On planet Earth, the region of Pangaea that would become North America always stayed north of what would become South America. In Planet Brain, the somatosensory cortex is in front of the visual areas no matter what. The arrangement of some of these key regions of our cortex is shown in Figure 1.5.

PLANET BRAIN: THE CITIES AND COMMUNICATION NETWORKS

The capitals of Earth interact with each other in different networks, sometimes associated with broad regions, such as Asia-Pacific or the European Union. But they can also overlap in their activities with capitals outside of those regions. Financial hubs such as Tokyo and New York engage in a constant flow of information exchange across half the diameter of the planet. The major hubs of the brain do this too.

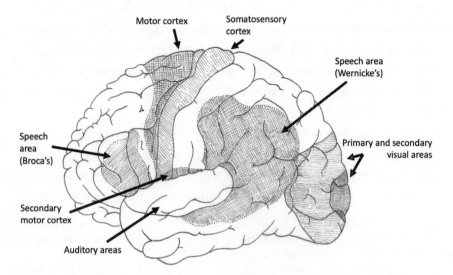

Figure 1.5. Some of the major countries of Planet Brain.

In the brain, we call these interacting hubs and the transportation routes that link them the "connectome."* And the hubs can specialize in more than one process. London, a hub of both academia and finance, interacts with some hubs in an academic network and others in a financial network. Similarly, the metropolises in our heads can participate in more than one group of connected hubs. Each of these groups of connected hubs is called a network. Here, we'll take a look at some of the key networks traversing Planet Brain, with special focus on a trio that are crucial to global and social cognition.

On Earth, it took us a long time, on the scale of thousands of years, to develop our transportation and communication hubs and the strong links among them. The mammalian brain required even longer, on the scale of millions of years, for its networks to evolve. Humans and other primates in particular have an abundance of long-range connections compared with other species, suggesting a more globalized brain.[5] It's probably not coincidental that these longer-range connections, many of which are specific to us alone, seem to support features that humans have uniquely taken to their current limits,† including tool use and language.[6]

When researchers looked at primates only, comparing connections within chimpanzee versus human brains, they found that the chimp has only three connections unique to its species. All of them link the two hemispheres of the brain, hallways between two rooms. Humans, by comparison, have many connections both between and within the hemispheres, linking two regions or multiple regions to each other, suggesting both global and local connectivity.[7] Other work has identified similar differences in connections between monkey or ape brains and ours.[8]

The development of hubs and connections during the evolution of the human brain also may have elaborated more ways for that brain to express itself to the limits of life function. Of all the species, we have the fewest constraints on our behavior and can take it to some surprising extremes without threatening

---

* In fact, the study of this connectome is called hodology, which means the study of roads or pathways; *hodos* = Greek for "paths."

† Other species use both tools and language, but we're the only species that has placed a helicopter on Mars.

the existence of the species.* Our overwhelming numbers and the infinite continuum of human uniqueness attest to this lack of constraint, despite ever-present efforts in some cultures to enforce narrow bounds of conformity.

But what, exactly, are these networks doing?

"I AM DEAD"

A fifty-three-year-old woman had moved from the Philippines to the United States only a month before being admitted to the psychiatric unit of a New York hospital. Her very worried family had dialed 911 after the woman, called "Ms. L," began to complain that she was dead.[9] And that was not all. She said that she could smell the death, an odor that clung like rotting flesh. Her conviction of being dead was so strong that she had asked her family to transport her to the morgue, where she could be with other dead people.

Hospital clinicians learned that she had notable symptoms of depression, including hopelessness, loss of appetite, sleeping a lot, and low energy. In fact, she had been taking an antidepressant for about eighteen years, although she could not tell them which one or how much of it. And, she disclosed, she was anxious that paramedics were trying to burn down the home she shared with her brother and a cousin.

A person who is convinced that they are dead—or have lost some part of their body or their soul—has a rare condition known as Cotard syndrome.† Not uncommonly, this syndrome co-occurs with other conditions, such as depression and psychosis. The delusions can lead to suicide attempts, and many people with Cotard stop eating, because, after all, if you are dead why would you need to eat? Given these exigencies, a measure of last resort has often been employed to fritz the conviction away: electroconvulsive therapy, or ECT.

In the case of Ms. L, the clinicians sought a different route. They started her on an antipsychotic drug and an antidepressant, given her evidence of depression with psychosis. She did not want to be treated at all, but her

---

* Individually, the risk that we'll experience terrible or fatal outcomes by going to extremes becomes more personal, of course.

† So-called after Jules Cotard, who first described it in 1882. The initial case he described involved a forty-three-year-old woman who was convinced she consisted only of skin and bone and would exist eternally if she were not burned.

family went to court to force treatment over her objections. After she was switched to a different antipsychotic, she had an episode that involved either fainting or a seizure. They switched her meds again and added an antianxiety drug to the mix. After about a month on a regimen of three drugs—antidepressant, antipsychotic, and anxiolytic—she was discharged from the hospital, saying she had no more paranoid fears or delusions that she was dead, and she "expressed hopefulness about her future."

Ms. L's experience sounds like something out of a gothic nightmare story, but her delusions and paranoia didn't come from a book. They arose in her brain.* The therapies that seem to have worked on her (and one they did not try, ECT) operate directly on the brain. But what do they do that yields improvement and a hopeful outcome?

A case study of a different patient, a teenager with brain inflammation and Cotard syndrome, showed that the young man had abnormalities in metabolism in two key brain areas: the front of the brain, known for its role in personality and decision-making, and another region involved in self-awareness. Studies in other patients with the syndrome had implicated these regions, along with a series of hubs running down the center, or midline, of the brain, where your hair parts in the middle.

These regions are nodes in networks that operate to define "self" for us, as distinct from not-self, or other people. They also may work together to cue us to our internal sensations and being alive. In the case of the teenager, after drug therapy failed to help him, ECT was performed, and he experienced complete remission.[10] The clinicians who treated him wrote that the multiple brain regions implicated on imaging set aside the idea that a single affected brain region could explain Cotard syndrome. Instead, a scattering of regions seemed to be involved.

The results of these imaging studies implicate brain networks and their dysfunction in this striking inability to sense one's own existence. The findings in Cotard point to problems with regions scattered throughout the cortex. Despite their distance, these hubs form networks that likely function to create our sense of self as distinguished from others, and to register our

---

* Indeed, her ailment showed a strong association with neurological diseases, such as pain syndromes, seizures, and strokes.

internal sensations, emotions, and pain (which people with this syndrome often do not sense).

THE POWER HOG

In the late 1970s, the first hints emerged that a large, power-hogging communications network of major hubs might be continually active in our brains. But it wasn't until 1997 that researchers noticed that some connected areas that went quiet during goal-directed tasks became active during rest. The tasks in these studies were "non-self-referential," meaning they were not personal, unless there's something personal in being tasked with listening to a string of numbers and repeating them later. At any rate, when a participant was lying still and spacing out, the regions that were silent during number recall showed signs of heightened energy use.[11] That raised the question of why, if the person wasn't doing anything?

In 2001, Marcus Raichle and his team published the first report, now often cited, specifically characterizing this mystery network that settled in quietly during a task but lit up with activity during rest. The paper was titled "A Default Mode of Brain Function" because the network seemed to dominate during a "default" state of just lying there.* The name, it turns out, perhaps wasn't quite apt: the network does a lot more than dominate during "rest," which, as we shall see, isn't really resting.

Regardless, the terrible and terribly dull name, despite a few halfhearted attempts at variation, has stuck. The network is usually called the default mode network, or DMN. It does a lot more than use energy while we're lying around with our eyes closed, and it's not the only network worthy of our attention.

That original paper heralding the existence of the DMN has, in the intervening twenty years, been cited in almost four thousand other papers, as of this writing. Clearly, Raichle and colleagues were on to something. One enduring takeaway was that a large part of the brain everyone thought was

---

* Raichle seems to have some regret about the title, noting in a 2015 review, "Parenthetically, it had not occurred to us that others would anoint the constellation of areas exhibiting this unique behavior as the brain's default mode network. The name obviously caught on." It did. Sorry—now it's all over this book.

basically silent was busily humming away, putting out to pasture the idea that we "only use 10 percent of our brains" at a time.*

As the stack of DMN-related findings piled up, so did new information about the network. It has subnetworks within it, each with a specific set of responsibilities. One operates in the front of our heads to take in sensory information and support emotional processing. Another lying right next to it helps us make judgments in relation to ourselves, and a postulated third is related to our ability to summon memories.[12]

The DMN has nodes and connections that cross each of these regions, and devotees† have called it the "backbone of cortical integration"[13]—a fancy way of saying that it connects and orchestrates the activity of important regions in the cortex.

The authors of a 2021 review of the DMN and its roles characterized it as the "sense-making network," which is so much easier to remember than the actual name. These authors argue that the DMN is always in the process of taking in information, meshing it with previously shaped personal narratives, and creating frameworks for understanding what's going on around us. In other words, it makes sense of the world and our place in it, something that people with Cotard syndrome seem to have become unable to do.[14] Unfortunately for those of us who read and write about the DMN, it is far too late to rename it.

THE CITIES OF LONDON, TOKYO, AND THE MEDIAL PREFRONTAL CORTEX

It would be great if we could name nodes of networks like the DMN after the capitals of the world because we could probably remember them more easily: "London, on the continent of Frontal Lobe, plays powerful roles in executive function and personality." But neuroscientists, in their infinite wisdom, instead have named these features based (usually) on their relative location in the brain.‡ To make matters more complicated, they don't all use the same names.

---

* An enduring myth. We use the entirety of our brains.

† Brain networks have fandoms.

‡ In their defense, we do this for all our organs and tissues when identifying their anatomical or functional location.

The lingo may take some getting used to, but the basics are that things are on the left and/or right (usually both) or in the middle (medial). They sometimes have a continental designation—their lobe—incorporated into the name. And just as we use north, south, east, and west to indicate broad direction, the brain has its own compass rose. The directions in the brain are anterior (frontward), posterior (behindward), dorsal (topward), ventral (downward), medial (down the middle), and lateral (to the side). Are you ready now for the quiz?

Where would you expect to find the medial prefrontal cortex?

That one's easy (maybe) because we learned at the beginning of the chapter that it's in the middle of the cortex in the most frontward region of our brains.* It is also a major node, or metropolis, in the DMN (Figure 1.6). If you traveled deep into the brain in a straight line from this node toward the back, or posterior, you'd end up visiting the great capital of posterior cingulate cortex, known to the enlightened as the PCC. Along the way, you'd have passed through a node of another circuit called the salience network (SN). This node is the anterior cingulate cortex. It sits in front of (anterior to) the PCC, another DMN node. This pair makes up the distant twin cities of the megalopolis cingulate cortex, which rests like a snug collar atop the corpus callosum (cingulate means "belt" or "band").

Figure 1.6 shows some of the nodes of the DMN and the salience network. You do not need to memorize all of them. There will not be a quiz. But take a moment to associate, where possible, the directional and lobe-related names to get a feel for some of the major cities of Planet Brain.

And now for a bit of a twist. These major cities of Planet Brain can participate together in different configurations. Despite the fixed appearance of the two-dimensional drawing below, these hubs don't always work in lock-step within a network. Some connections are steady—perhaps not unyieldingly so, but at least consistently connected. Others are more flexible, able to shift or shut down in specific situations. And in some cases, nodes will

---

* For extra credit: the medial prefrontal cortex is divided into two functional regions, one on top of the other. If your only word bank is anatomical terms, what would you call these? If you said, "Dorsal medial prefrontal cortex on the top, and ventral medial prefrontal cortex below it," you are right.

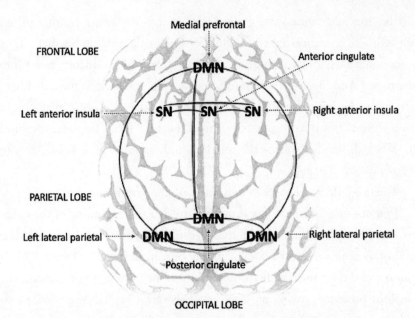

Figure 1.6. A view from the top of some components of the default mode network (DMN), including the medial prefrontal area. In this image, the DMN is shown crossing paths with the salience network, which is responsible for sorting out signal from incoming noise, or identifying what is salient in the moment. After Van Ettinger-Veenstra et al. 2019.

only transiently take part in a network, dipping in on rare occasions for just a bit, like a celebrity making an appearance at a gathering that's important for their career.[15]

DON'T LET THE DMN FOOL YOU INTO THINKING IT'S THE ONLY NETWORK ON THE PLANET

Scientists used to focus on individual regions of the brain because of technical limitations, but as technology has picked up pace, ambitions in brain studies have followed. Instead of poking around one region at a time, researchers have turned to an overview of entire systems, networks like the DMN that span the cortex, united in function, connected by interacting hubs, and interacting with one another. We are, I think, midway along the enthusiasm curve for these networks and their interactions, and how we think about the brain will probably evolve in the future just as it has in the past. But even if networks such as the DMN take on new contours, they still serve as metaphors for what's happening in our brains. Future discoveries may shift the weight of their relevance in function or form, but they remain an excellent symbolic shorthand for how our cognitions can be distinct yet overlap.

Throughout this book, we (and our DMNs) will bump into a few other networks that help us define different functions of our brains.* Among the ones we will encounter most is the salience network, which sorts out signal from all the sensory noise around us. The hope is that the network sorts accurately and doesn't omit important information or overemphasize the irrelevant, both of which are possible.

Two key nodes of this network are the anterior insulae, one in the left hemisphere and one in the right (Figure 1.6). If you could pry the frontal and temporal lobes apart at the sulcus that separates them and lift each one, you'd find the insulae tucked underneath. Each anterior insula lies toward the front of your head.† A third important hub in the salience network is one we've already met: the dorsal anterior cingulate cortex. This central hub lies on top of (dorsal‡) and wraps around (cingulate = belt) the front (anterior) part of the corpus callosum, or tough body, that connects the two hemispheres. All three of these "metropolises" connect to structures beneath the cortex, including the amygdala, the one that processes emotion.

In part thanks to our salience network, we can register important things and make instinctive decisions about them, such as walking around the dog poop some rude neighbor left on the sidewalk. During that walk, we also can monitor the sky for an approaching storm and tailor our route based on how quickly the clouds are approaching.

Perhaps on that stroll, we are also contemplating a difficult problem. As we turn it over in our head while walking, stepping around the dog poop, and keeping a close eye on the clouds, we are engaging our central executive network, or CEN. Because nothing in the brain can have just one name,

---

* To keep things straightforward, examples in this book will often use three networks in particular, but they are not the only functional networks that have been identified, and what's known about their contours is always being refined.

† And yes, there is a posterior insula, the part lying horizontally toward the back of your head.

‡ Like a dorsal fin. The opposite is ventral, which means "belly."

this is also known as the fronto-parietal network, thanks to the location of the main nodes (frontal, parietal) that form it.*

The CEN sits in its corner office with ceiling-to-floor glass windows and a view onto all the components of the problem at hand. With it, we sort out what's relevant and work on steps to a solution. For example, neighbors who let their dogs poop on the sidewalk every day are a problem. With the fresh memory of having just skirted around the canid-sourced solid waste and your eye on the advancing clouds, you can step through a process of asking the neighbor to cut it out without starting a feud.

The salience network may operate as a sort of toggle between our DMN and our CEN.† The back and forth may help us settle on an optimal social approach to the neighbor relative to our self-need to keep the peace.

Through our sensorimotor network, we take in information that our salience network filters. We saw (sensori-) the dog poop and stepped (motor) around it. We see the clouds edging toward us, and perhaps feel a raindrop or two. In response (motor), we pick up our pace to get home before the drops turn into a deluge. The neighbor can wait. For now.

This information is in turn coordinated between the DMN (self-interested or self-involved factors in play) and the CEN (slots the information into the current scheme and tackles what should be done about it). We respond with what was decided, sending the decision back out as a motor response, in this case, hustling homeward. The more our salience or executive networks are engaged, usually the more backgrounded the DMN is. In other words, the more we focus outward, the less the DMN draws us inward.

This triad of networks is one you'll see mentioned often in the chapters that follow. The DMN and CEN are thought to anticorrelate, which means that when one is revved, the other is muffled. The salience network, possibly acting as the toggle between self-focused thought and achieving a goal,

---

* To make this even more complicated, it also has been characterized as a division between the CEN and the dorsal attention network, or DAN. If any field needed a central body making calls about terminology to limit confusion, that field is neuroscience.

† These networks have different names depending on who's authoring the studies, so I'm going here with the ones that seem to be common and that make sense based on what the network does. The exception, of course, is the DMN, with the name that doesn't tell us much of anything *and* is misleading.

works the switches between them. Together, the three networks may operate in a triumvirate to keep our shoes poop free, our heads dry, and our neighborhoods free of strife.[16] That's our brain, including our social brain, in action.

CITIZENS BRAIN

The citizens of Planet Brain are a motley bunch, roughly divided into "neurons" and "not neurons." The neuronal citizens are the bodies electric of the brain, signaling to one another at lightning-fast speed without even touching. But they couldn't do it without all the other residents of Planet Brain, which collectively are as numerous as the neurons. Just how numerous is that?

In 2013, neuroscientist Suzana Herculano-Houzel gave a TED Talk titled "What Is So Special About the Human Brain?"[17] It could just as readily have been called "Brain Soup." Herculano-Houzel is to "blame" for the fact that we lost about fourteen billion from our brain's neuron count. Her work with "brain soup" brushed aside the oft-claimed count of one hundred billion (which she herself had cited before, noting that she could never track down the original reference for it*). Math (a hundred billion minus fourteen billion) tells you that the number her group came up with was eighty-six billion. How did they go about counting high enough to arrive at such a specific reduction in actual versus mythological neuron numbers?

Obviously, counting neurons that number in the billions one by one is not tenable; nobody's got the time for it (literally—it would take about ninety-five years by one estimate to count just to one billion).[18] But Herculano-Houzel landed on the idea of making brain soup to count cells. Once you read about her strategy, you'll likely experience two reactions: amazement at how straightforward it seems, and disgust because it sounds disgusting. Your insula is giving you an assist on this.

To make brain soup, you need a brain (or part of one), detergent, an antibody (an immune system protein) that targets another protein present in most neurons and only in neurons, and a fluorescent tag you can attach to that antibody. First, the tissue gets a dunk in formaldehyde, which fixes the

---

* As one researcher pointed out to me, it's possible that people were just rounding to the nearest hundred billion.

components in place. The next step is to dissolve the brain in the detergent, which breaks up the fatty cell membranes—much like detergent breaks up grease—releasing the cell contents but leaving the nuclei intact. Now you've got a "soup" full of free nuclei from the brain you dissolved, something Herculano-Houzel compared to hazy apple juice. There's the disgust.

To count neurons in this concoction, you take samples of the soup. To ensure an even distribution in your samples, shake up the soup a bit to evenly disperse the contents, like stirring real soup before ladling it to get a nice allocation of its components.*

Incubate these samples with that antibody that sticks only to neurons, add the fluorescent tag that sticks only to that antibody, and then look at a bit of the "soup" under a microscope and count what you see. Everywhere there's a fluorescent glow, the antibodies will have piled up only on a neuron's nucleus. They won't have attached to other cell nuclei in the brain. A machine can count the glows for you. Once you've got your nucleus count, you just correct it for the full volume of the brain soup, and voila! You've counted billions of neurons.

When Herculano-Houzel and her team did this in their 2009 study with human brains, they came up with eighty-six billion neurons, not a hundred billion. The difference, as Herculano-Houzel pointed out in a Vanderbilt University profile of her work, is an "entire baboon brain and change."[19]

THE BODIES ELECTRIC

Those eighty-six billion neuronal citizens of Planet Brain are not all alike. The brain has dozens of types of neurons, and we aren't going to go over them here. Throughout the book, where relevant, I will specify a type if it matters.

One reason these cells can be so tough to categorize is that we've got a half dozen ways to do it. Some neurons bring in information from the senses, and some neurons send out messages about how to respond, so we categorize them that way. But some neurons inhibit the cell they communicate with, whereas others excite it, and we categorize them that way.

---

* Apologies if this ruins apple juice and/or soup for you.

Some neurons send out their projections, called axons, to the same target, so we categorize them based on that behavior. Some neurons have a single cell body, an axon, and a terminus, whereas others have a cell body at the center of two axons sticking out in different directions, like they're doing the splits. Still others have many branches around the cell body—called dendrites because they look like naked tree branches (dendrite = "branched")—whereas others have no branches at all. Neurons also can be characterized biochemically, based on which neurotransmitters (neural-signaling molecules) they send or receive.

It's complicated. But Figure 1.7 is a diagram of a simple neuron so that you'll have an idea of the appearance and basic anatomy of the average neuronal citizen of the brain. Its shape just screams, "Electricity!," which is appropriate because that's what it uses internally to transmit a signal.

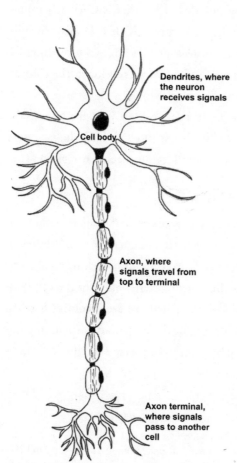

Neurons receive incoming signals at the cell body, through the dendrites, which look a bit like a fright wig. Dendrites in turn have twig-like extensions called spines that take these signals. If an incoming signal is of the "excitable" sort, it can build up where the body and the axon connect. When that buildup reaches a certain threshold, there's a burst of voltage pushing a current that travels down the axon to its terminal. That's the neuron "firing."

At the terminal of the axon, this wave of electricity can

Dendrites, where the neuron receives signals

Cell body

Axon, where signals travel from top to terminal

Axon terminal, where signals pass to another cell

Figure 1.7. A basic neuron.

trigger a release of chemicals into the space outside the neuron, called the synapse. These chemicals diffuse to the next cell, bind to the dendritic spines there, and trigger a voltage change in the neighbor.

In sequences like this,* neurons pass along messages. Their axons, often bundled, form the communication networks that link the metropolitan hubs of the connectome. Unlike the human citizens of planet Earth, these citizens of the brain are also structural and functional contributors to their planet.

The dendrites that make up that fright wig around the cell may have many, many spines, through which a neuron can receive thousands of inputs from other neurons. Two key signaling chemicals are involved in this process. One is glutamate (aka GLUT, pronounced "glute"), which is associated with "exciting" or activating neurons. The other is GABA (gamma-aminobutyric acid, which isn't as fun a name), which is associated with inhibiting or dampening neuronal enthusiasm.† If the incoming signal is inhibitory (GABA), then the neuron doesn't build up tension at the connection between cell body and axon and doesn't fire.‡

For a real-world example of how these two work on your brain, consider this: consuming alcohol promotes a GABA boost and dials back the excitatory glutamate, leaving you with that loose, relaxed feeling alcohol brings on.

THE GLUE

The neuronal citizens of Planet Brain are only about half of its resident population. The other half are non-neuronal cells, or glia. The antibodies in brain soup would have ignored these cells. You might have learned at some point in your schooling that the brain's grey matter consists of the cell bodies of neurons, unprotected parts of the cell that receive incoming signals. You'll also find a lot of other components in the grey matter, including naked axons and glia (which means glue—they once were thought to be connective tissue), among other things.

---

* It's not the only way neurons signal to each other, but it's a good basic construct.

† Mnemonic: GLUE = GLUtamate Excites; GABI = GABA Inhibits.

‡ The brain relies on a host of other signaling molecules that play various roles in communication between neurons. These neurotransmitters include serotonin, dopamine, and noradrenaline. As with the networks, for the sake of focus, I have tried to select representative examples of processes rather than providing a compendium of action for each of these.

The brain's white matter consists of axons that have been wrapped in the fatty protective insulating sheath that makes them glisten white, giving them their collective name. According to one estimate, if you separated out these eighty-six billion fibers and laid them end to end, they would stretch across more than a hundred thousand kilometers, or more than sixty-two thousand miles. That's just from one human brain. That's quite an extension cord, and it is what makes the connectome possible.[20]

Glia called oligodendrocytes produce the myelin insulation for these axons. Another type of glial cell, the astrocyte, has many roles, one of which is to ensure energy supply to the neurons, passing nutrition to them on demand.*

THE SOCIAL PLANET

The backdrop of evolution in this chapter can serve as a reminder of something: the unit that evolves in nature is not the individual but the population.† Our brains, collectively and individually, are the outcome of millions of brains that preceded us. Each of us has a brain that's just part of chains of brains shaped over millennia into what we have today.

The pressures that shaped those brains weren't individual either. They, too, were collective. Our interactions with each other—as mate to mate, parent to child, sibling to sibling, group to group—molded our brains into organs that are responsive to these sculpting factors.

Yet when we pick up a book about making our brains better, the focus is almost always on the individual. *You* can hack your brain to get what *you* want from *your* brain, to make *your* brain better for *you*. As though that brain cropped up like a novel plant after a fresh rain, no prelude, no current company, no sequel.

I propose that we take a more collective approach to this tailoring. Rather than each of our brains being a lonely planet, they are connected and

---

* This description of glia is extremely basic. They are a complex group of non-neuronal cells, and the list of their roles and their importance just keeps growing. That information could fill a book. For this book, I just want you to know that they exist and have these roles.

† The definition of "evolution" is a change in the frequency of a gene variant in a population over time.

interactive and in exchanges with hundreds or thousands of other brains, influencing each other along diverse pathways, like neurons themselves. Individualism is all well and good, to a point, but almost none of us live alone, in a cave, cut off from the rest of our species.

Our brains do not stop at the membranes that encase them. When we seek improvements—when we strive to tailor the brain—we should always keep in mind that with rare exceptions, we are tailoring *brains*. Not *a* brain. Indeed, I wrote this book, and although we (likely*) have not met, I am sharing the work of my brain with your brain in the hope that our collective brains will all end up at least a smidge better than when we started.

END-OF-CHAPTER CHECKLIST

✓ Planet = brain
✓ Continents = lobes (frontal, parietal, temporal, occipital)
✓ Countries = association areas and somatosensory and motor cortexes
✓ Cites and communications network = hubs (such as the left insula or the posterior cingulate cortex) in networks (such as the DMN)
✓ Citizens include neurons and glia
✓ Communications = axons, synapses, dendrites
✓ Taken together = you (singular) and all of us (plural)

---

* If we have met, well then, hello again!

CHAPTER 2

BRAIN TINKERING: TOOLS AND TECHNIQUES

H ELLO! WELCOME TO the chapter on interventions—what they are and the reasons, if any, that they are expected to work. Because much of the evidence in this book relates directly to the use of these techniques in studies, this chapter also takes you on a tour of the ostensibly rational world of scientific research. As you'll find, it may be rational sometimes, but it's not always or even usually transparent. And often it's not coherent, because in neuroscience the questions may be similar, but how researchers go about answering them is all over the (brain) map. It's hard to get a good idea of what findings mean when they are always "mixed."

As we've already seen, even declarations with no evidence base at all can become dogma. That means it's up to you to develop a discerning mind when it comes to assertions about the brain, and you'll find some guidance for that here too. With this information in hand, you'll have the necessary tools to consider big claims about interventions before you ever consider tooling around with your brain.

Poking around in our brains has a long and only partially honorable history. We've been trying to zap them, attempting to relieve them of various imagined buildups, and misinterpreting their manifestations for millennia.[1] Even now, in the twenty-first century, myths that persisted for decades or more are crumbling as fancy technologies give us increasingly specific and targeted ways to see into and futz around with our brains without actually touching them. I guess, at least, it's good that we've moved beyond trepanation.*

In 1995, *Newsweek* was a print magazine that lots of people paid to read. I know that sounds archaic and unfamiliar, but it's true. That year, people could buy the March 27 issue at the newsstand for $2.95 and find out all about "THE NEW SCIENCE OF THE BRAIN." In smaller type on the cover, we were promised that we'd learn "Why Men and Women Think Differently."†

Inside, seven full pages of a feature story and assorted sidebars covered cutting-edge research on the brain and how it could confirm all our existing biases about male and female sex dichotomies. The opening lines of the feature story‡ read, "Of course men and women are different. *Boy*, are they different. In every sphere of life, it seems, the sexes act, react or perform differently. Toys? A little girl daintily sets up her dolls, plastic cups and saucers, while her brother assembles his Legos into a gun and ambushes the tea party."§ After a couple of other similar examples applied to adults, it continues, "Stereotypes? Maybe—but as generalizations they have a large enough kernel of truth that scientists, like everyone else, suspect there's *something* going on here."

All of which sets up the article's intro to a pair of "new technologies" known as functional magnetic resonance imaging (fMRI) and positron

---

* An ancient and sometimes surprisingly effective surgical procedure that involves making a hole in the skull, possibly to relieve pressure or address a head injury.

† And to give you a sense of how the times have and have not changed, a corner headline read, "Affirmative Action: Playing the Gender Card."

‡ Written in the contemporary context by the late and highly respected science journalist Sharon Begley, who did not own a time machine. Most of us could not readily withstand (or stand) a review of our writing from a quarter century ago.

§ Oddly enough, despite my sex, reading this made me want to ambush something too.

emission tomography (PET), both of which "catch brains in the very act of cogitating, feeling and remembering." The results of one study, we are told, show differences between men and women when their brains are "idling." In men's "idling brains," a system that's part of the "reptilian brain" showed higher glucose uptake on PET in a "primitive region" that controls "highly unsubtle expressions of emotion, such as fighting."* Meanwhile, in the "idling" brains of women, glucose uptake was heightened in the posterior cingulate gyrus,† which the article described as an "evolutionarily newer addition to mammals' brains." The study had thirty-seven men and twenty-four women. Of these, thirteen of the men (a third) showed the latter pattern, and four of the women (a sixth) showed the "primitive region" pattern.

About such results, the article quotes a researcher who noted that "some of the women's brains looked like the men's [which] is true of all these sex studies." Indeed. In one study the article describes, *42 percent* of the women had limited emotional responses to nonsense words, yet the conclusion was that women tend to have more emotional responses to nonsense words than do men. So maybe there wasn't really that much of *"something"* going on.

The article mentions some brain claims that later revelations mothballed, including the concept of the "reptilian brain" and the idea that we can be more "left-brained" or "right-brained." Like the myth that we "only use 10 percent of our brains" or the idea of the "triune brain,"‡ which is related to the "reptilian brain," targeted investigations have found no basis for these beliefs.

---

* To be clear, fighting is an extremely complex behavior that is not reducible to emotions, plural, much less a single, uncomplicated emotion, and people of all sexes fight.

† We've already met this structure as the posterior cingulate cortex, which includes this gyrus and its adjacent sulcus. It's the ceiling of the limbic system and the back end of the "belt" that lies on top of the corpus callosum. The terminological inconsistencies and overlaps in neuroscience can be maddening.

‡ This notion was intended to reflect the inferred evolutionary steps of making a brain like ours. The three parts that put the "tri-" in "triune," in order of their appearance in the family tree, were proposed to have been the reptilian complex (aka lizard brain), the limbic system, and the "neomammalian complex." We now know, for example, that the thalamus, part of the limbic system, hasn't just sat there being thalamically unchanged through millennia but instead has undergone adaptations along with the cortex.

People in 1995—journalists and scientists included—were not consciously colluding to support the patriarchy or reinforce gender stereotypes. But they and you and I cannot help but carry our biases with us wherever we go. How else can you interpret results as revealing something distinctive "about women" when almost half the women among the small number included in the study responded like the men?

Even our mood and attitude can affect how we ask research questions, answer them, and interpret those answers. We capture this most simply when we talk about a glass half full or empty. Do you view it as half full, you bright-eyed optimist? Or as half empty, you dour downer? In the conduct of scientific endeavor,* we're supposed to just look at it and say, "That's half a glass of milk." Being human, we almost never leave it at that.

These biases matter. Confirmation bias means that we'll scan information and home in on the bits that confirm what we already think, sweeping away the rest. I try to avoid doing that in presenting information in this book, but I am no robot, and neither are you. The best we can do is consistently remind ourselves of this fallibility and as consistently force ourselves to confront it. Our skill at shedding assumptions is directly linked to our skill at truly attending to what the data say—all of it, not just the bits we like. Salience network, I'm talking to you.

Publications in science are riddled with layers of bias, from their origins to their interpretations. Knowing who's asking the question and why is crucial to understanding how that question is approached and answered, as the next chapter illustrates. And one key source of bias that may elude detection is publication bias itself.

STUDY? WHAT STUDY?

Researchers (or companies trying to sell you things) may be emotionally and financially committed to positive results from a study. If they don't get the results they wanted—well, there's not really a commandment decreeing that they have to publish the findings. In fact, it's notoriously difficult to get many journal publishers interested in outcomes that show no differences or

---

* Or, as we shall see, of mindfulness.

effect. That's unfortunate, because these so-called negative results, scientifically speaking, are just as important as positive ones. The upshot of these sins of omission is that we see a huge proportion of the positive findings and a much smaller percentage of the negative ones.

How should we interpret positive findings when negative results can simply be suppressed? For trials involving humans, researchers can register studies publicly before starting them. These registrations are attached to unique numbers that anyone can search online in the relevant trials database.* But that doesn't work very well either—it just tells us how big the problem is. The authors of a 2021 editorial in the *New England Journal of Medicine* wrote that during the SARS-CoV-2 pandemic, of all the completed or terminated trials posted to the site, only 8 percent had also posted the trial findings. That leaves 92 out of every 100 completed or terminated trials with unknown outcomes. Obviously, what is made public is biased.[2]

If a trial has been registered for years but no publications have emerged from it, the radio silence could suggest that the researchers got nothing of interest from the work. Sure, they didn't publish that result, but at least there's a record that something was planned.

Not much, is it?

Some journals add another layer by requiring authors reporting trial findings to follow a specific rubric and to pinky swear† that they're being fully transparent. For example, the CONSORT (Consolidated Standards of Reporting Trials) guidelines cover randomized trials, the STROBE (Strengthening the Reporting of Observational Studies in Epidemiology) guidelines cover reporting of observational studies, and the PRISMA (Preferred Reporting Items for Systematic Reviews and Meta-Analyses) guidelines cover how researchers should conduct and report their findings from systematic reviews/meta-analyses. At the end of this chapter, I briefly cover a few types of studies if these phrases don't mean much to you right now.

Finally, beware of the salami slicers, the researchers who conduct a single study and then incrementally publish bits and pieces of the results in

---

* Go to clinicaltrials.gov, input "ketamine" into the "Other terms" search window, and take a gander at some of the thousand-plus trials listed there.

† Not really, but it's implied.

different papers without disclosing that they all come from a single study. Why? Padding. If you're trying to sell your idea (or your product) and you want to convince evidence-seeking buyers that it's legit, would you want to point them to a single publication with the results from the only trial you've conducted? Or would you want to enumerate a dozen or more articles, still all from one study, but giving the impression that each represents a unique study with unique findings that build your case? As you'll see in the chapter on cognitive interventions, detecting who's chosen which route can be important to how you receive their claims.

The passage of time can bring clarity. More than a quarter of a century has elapsed since the publication of that *Newsweek* article and the often mythical neuroscience it cites. In the meantime, fMRI and other imaging techniques have blossomed into something like maturity, even as other approaches have sprouted up alongside them. With these new ways of looking into our same-old brains, we also have developed new interventions targeting what we find. Below, we'll take a look at some of these techniques and interventions to lay a groundwork of understanding what comes up in the following chapters.

And a word of caution: Alvaro Pascual-Leone, a Harvard neurologist who works with patients using one of the newer brain-stimulation interventions, told me that these technologies certainly offer "really powerful opportunities to understand the brain. . . . It is an amazing thing that we can actually gain an understanding of why they are benefiting patients today rather than twenty generations from now."

Yet, he said, that very appeal and potential to be effective are what make them dangerous. Our human facility at rapidly pushing inventions at a cutting edge means that sometimes, other considerations trail behind. Some of the promised interventions have found their way, for example, into the do-it-yourself realm before we even know what they're doing to yourself. "It requires a careful, appropriate approach," Pascual-Leone said, in part because "the potential is there, and we're only scratching the surface of that potential." It's a bad idea to unleash the kraken before we understand how to control it, but we often do so anyway.

THAT DOESN'T MEAN YOU LOVE BEER
AS MUCH AS YOUR SPOUSE

One of the imaging techniques that offers us exquisite views of our brains is magnetic resonance imaging, or MRI. This method relies on the content of water's hydrogen atoms, or their protons,* in tissues. A person lies inside the tunnel of a giant magnet, which when activated causes these protons to align in the same direction. Then the hydrogen ions are hit with some radio waves, which bobble them out of alignment. When the radio waves stop, the ions spring back into alignment, releasing radio waves themselves at the same frequency, or resonating. Receivers can pick up these waves and use their different frequencies to create a greyscale image of the targeted part of the body. Differences in water content,† and thus proton content and resonance, underlie differences in the returned radio waves.

The resulting pattern creates an image of the tissues, varying from white to pitch black, depending on the proton (and water) content. For example, in one specific MRI mode, the watery cerebrospinal fluid that fills the ventricles, or large spaces, of the brain—the oceans of Planet Brain—show up as bright white on MRI. Areas of injury where fluid has accumulated will also show up brighter than the healthy background tissue in the same imaging mode.

But these are still images that don't give hints of where the brain shifts activity depending on what it's thinking. Another kind of MRI, functional MRI, has been developed to take care of that. To add the "functional" part, this imaging approach takes advantage of hemoglobin's relationship with oxygen. Neurons that increase their activity need more oxygen, and blood is responsible for making the delivery. In the blood, oxygen attaches to a compound known as hemoglobin, which resides in the red blood cells. Hemoglobin clutching its oxygen responds to a magnet differently from unburdened hemoglobin, and fMRI can pick up this difference, showing in real time when the blood starts to carry more oxygen to specific parts of the brain.

---

* A hydrogen atom consists of just a proton and an electron. Taking the electron away, or ionizing it, leaves only the proton, so hydrogen ions, H+, often are just called protons.

† Water has two hydrogen atoms.

What's being detected is called the BOLD—or blood-oxygen-level-dependent—signal. With fMRI, researchers can give the person in the MRI machine a specific task and then capture what happens to blood oxygen levels across Planet Brain. If specific regions show an increased BOLD signal, the implication is that the brain kicked up local activity during the task.

A subtype of fMRI study, resting-state fMRI, involves lying in the scanner doing nothing. It was in this way that the DMN and other networks first made themselves known. Although as described in Chapter 1, Raichle and colleagues are credited with characterizing the DMN, earlier hints of such resting activity and use of resting-state fMRI were published in 1995,* in the work of Bharat Biswal[†] and coauthors at the Medical College of Wisconsin.[3]

But the technique isn't perfect.[‡] Cognitive neuropsychopharmacologist Manoj Doss, a postdoctoral research fellow at the Center for Psychedelic and Consciousness Research at Johns Hopkins University, told me that in a lot of experiments using fMRI, the "task" part, especially if it's a resting-state protocol, can be "really unconstrained." Imagine lying in a tube, being subjected to the racket of the MRI machine—we're talking *crashcrashcrashcrashcrash, bangbangbangbangbang, beeeeepbeeeeepbeeeepbeeeepbeeep*, at volumes so high that most people wear headphones or earplugs while inside the contraption. Plus, the tube is small, the ceiling just inches from your face. And you've been told to lie still and avoid focusing on anything.

"When people are just sitting there doing whatever the hell they want . . . they're gonna get some kind of ego dissolution, so that could be one thing happening," said Doss, referring to just letting the mind wander away from

---

* But after the *Newsweek* article.

† Now a Distinguished Professor of Bio-Medical Engineering at the New Jersey Institute of Technology.

‡ An extreme illustration of how misleading fMRI results can be if they are not analyzed properly is the infamous "dead salmon" study. The researchers imaged a dead salmon, and when they purposely did not follow proper steps to remove the chance of a false positive, their results suggested that the salmon, dead as it was, was engaging in a "perspective-taking task" while lying in the MRI machine. The whole thing is described here: http://prefrontal.org/files/posters/Bennett-Salmon-2009.pdf. It was so brilliantly done that it earned an Ig Nobel Prize, which "honors achievements in improbable research," in 2012.

thinking about the self. "But then another set of subjects are paranoid as hell in the scanner and are trying to get weird executive control."*

His point is that what people think about in the scanner in the absence of a specific task can be almost anything. To deal with this, he said, studies involving the resting state should usually include task-based assessments, too—a control task and an intervention task—to see what happens when participants all have a consistent assignment and aren't told to just let their minds wander randomly.

In addition to frequently using small, mixed groups of people, instead of larger groups of people who are similar in some way, such as sharing a diagnosis, fMRI-based studies suffer from a number of other issues. Doss ticked off a few: motion artifacts (people don't stay as still as they should), artifacts from breathing, and pulsating of the brain itself, as we learned from the Egyptians. There are ways to control for some of these factors, but they still pose problems.

Jens Foell, a neuropsychologist and science communicator and previously an associate researcher in the department of psychology at Florida State University,† told me that one big drawback of fMRI studies is that the results are "strictly correlational." That is, one thing tracks mathematically with another, without a causal connection established between them. "You can look at [the results] to see what's going on in the brain, but that's all there is to it," he said. One pitfall is taking a similar fMRI result between two conditions as reflecting some common cause when the cause is not established. As an example, he recalls a study showing that people have BOLD signal increase in the same area of the brain when they are looking at a photo of a loved one as they do when viewing a bottle of beer. "From a neuroscience perspective, it makes sense because the area of desire processes in the same way," he said. "But that doesn't mean that you love beer as much as your spouse. It's very easy to overinterpret the meaning."

He also noted that the absolute level of blood oxygen is not informative. All the fMRI is really showing is changes in those levels. The challenge, he

* College students apparently tend to think a lot about sex.
† His Twitter handle is @fMRI_guy.

said, is determining what's background versus actual signal, given that the 10 percent brain-use myth is nonsense and the entire brain is using oxygen all the time.

Why get into all of this? Because much of the evidence of networks—the hubs interconnected like far-flung, communicating metropolises—relies on fMRI studies. Foell said that people get really excited about using fMRI to identify BOLD increases in a network, when in fact the study participants are just lying in the scanner, "doing nothing." The best findings that distinguish the DMN and other networks come from comparisons between people with diagnosed brain conditions and those unaffected by them. These comparisons sort out what looks like typical from atypical activation, an indirect way of confirming that the networks are real . . . and can be disrupted. Foell said that across a lot of studies, the DMN "was the network that showed up again and again."

To consolidate these findings, Foell said, echoing Doss, task-based research results are stronger than those based on participants who are directed to simply lie still, for the same reasons that Doss cited: the brain activity of people who are "off task" (i.e., doing nothing) is not uniform. You'll see similar critiques when it comes to tests of brain interventions and how the type of "control" condition can affect whether or not differences show up.

This messiness does not mean that fMRI data should all be thrown out the window, which is good because there is a lot of it. "I think that there are researchers who are pretty cynical who say that this should all go into the bin," said Foell. "It has its place, but you have to be careful with it."

RECORDING THE ELECTRIC BRAIN ORCHESTRA

The great benefit of fMRI is its high spatial resolution. But this technique captures moments *in* time and not activity *through* time, so researchers often add a method that captures the time element: electroencephalography, or EEG.

You'll recall from Chapter 1, I hope, that if a neuron is excited enough, it will "fire" and send a current down its axon, toward a target cell. If the neurons in a given area are collectively firing in a rhythm, they can create waves of firing and relaxing, firing and relaxing. When their signaling rises and falls in a pattern, the oscillations can be slow or fast, changing in frequency depending on our behavior. We can detect the rhythms with basic medical tools.

All of this is happening in your head, at different intensities depending on what you're doing and feeling. Disruptions in the oscillations can reveal a problem, such as a seizure. If you have ever had an EEG, you've had the pattern of your oscillations recorded on a piece of paper that looks like an earthquake readout.*

That brains produced waves of electrical activity became clear thanks to experiments on animals in the late nineteenth century. Physician and physiologist Adolf Beck first detected the oscillations in frogs, then demonstrated them in paralyzed dogs and rabbits. He showed that whatever established the pace of the waves, it was not heart rate or breathing. The rhythms seemed to be spontaneous, but applying a stimulus to specific regions of the brain could change their power. He was really demonstrating a direct effect on neuronal activity of a sensory stimulus, like touch, a hand clap, or a flashing light, without the actual presence of the stimulus.[4]

When Beck published his findings in 1891, European researchers had a tizzy, and some desperately tried to stake a claim to having gotten there first. Vasili Y. Danilevsky, a researcher at Kharkiv University, in what is now Ukraine but was then Russia, had written but not published a doctoral thesis describing similar findings in the dog brain. Beck, by publishing first, had simply beaten Danilevsky, a classmate of Vladimir Ulyanov (aka Vladimir Lenin), in getting the word out.

What neither of them knew—and no one else seemed to notice—was that yet another physiologist, Richard Caton in Liverpool, England, had gotten to press before anyone. He'd published a short description of spontaneous oscillations in monkeys and rabbits in 1875, and then a longer version of his findings in 1877, both in the *British Medical Journal*. Unlike Beck's report of the spontaneously oscillating brain responding to a stimulus, Caton's paper produced "no single ripple in the pool of physiologists."[5] Nevertheless, history's eye for the truth singles him out as probably having described the

---

* Which also shows waves that have been detected.

world's first EEG. He eventually became famous for being the Lord Mayor of Liverpool, suggesting that his career took a bit of a turn.

The first person to try to capture these oscillations in humans, however, was German psychiatrist Hans Berger. He was injured in a fall from a cavalry horse, and his sister, sensing that something terrible had happened, sent him a telegram. Based on this experience, Berger became convinced that telepathy was real (as you'll see later, we may well be on our way to making it so).[6] He abandoned a plan to study astronomy and turned instead to electrophysiology.

In his zeal to measure human brains, Berger used his children as apparently unwilling research participants. But his real breakthrough came in 1924 when he performed the first true EEG on a human, a seventeen-year-old boy undergoing neurosurgery. Berger called the two different spontaneous oscillation frequencies he'd identified "alpha" and "beta," a game-changing discovery in the world of neurology.[7] Both Beck and Berger died by suicide under very different circumstances. Beck died in the Janowska concentration camp, where the Nazis had imprisoned him for being Jewish. Berger took his life in the clinic in Jena, Germany, where he worked, after he'd participated in the regime's forced sterilization program. Perhaps this low bar he set for himself was what made him comfortable experimenting on his own children.

In a few of the chapters to follow, we will encounter the modern version of EEGs. We'll also encounter brain waves, the oscillations that Caton first detected 150 years ago. EEGs measure these oscillations somewhat crudely through the scalp. To pick up the brain waves, which arise from neurons in a region acting collectively, electrodes are attached to the scalp using a water-soluble glue, or the person can wear a cap of electrodes. The electrodes measure brain waves in broad areas, mostly designated by the continents of the brain, the lobes. So they'll register activity in the parietal lobe, or maybe at the border of the parietal and occipital lobes, and so on.

Just what are these waves?

THE MUSIC OF THE MIND

When neurons synchronize their activity and fire and rest in waves, the waves are called oscillations because they ebb and flow, up and down, up and down.

The brain waves have names, taken from the letters of the Greek alphabet, a trend that Berger kicked off. Here they are, in order of frequency range,* or cycles per second: delta (2–4 Hz), theta (4–8 Hz), alpha (8–13 Hz), beta (13–25 Hz), and gamma (>25–150 Hz). I remember their order as David Took All Ben's Gobstoppers, but choose the mnemonic that works for you. These days, we can measure the waves through the scalp, without being invasive, using EEG or an even more precise tool, magnetoencephalography, MEG.

These oscillations can encounter each other and interact like expanding ripples on a pond, so don't get the idea that your brain is generating only alpha waves one minute and delta the next. They play their music together. In fact, combinations of the oscillations may add depth and richness to what our brains can do, and their patterns are linked to specific tasks and

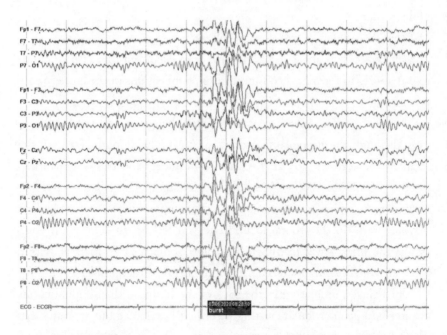

Figure 2.1. An EEG from a patient with epilepsy showing "spikes" (pointier) and "waves" (less pointy) of increased amplitude. The letter-number combinations indicate specific electrode placement, with the letters indicating the lobe (F, frontal; T, temporal; C, central, or the sulcus; P, parietal; and O, occipital). The electrodes are listed in pairs (e.g., Fp1-F7), and the voltage difference between them is measured.

---

* These ranges vary slightly across various publications.

behaviors.[8] An EEG can show all of them combined and can also sort out each signal from the overall noise, like breaking up the musical notes in a series of chords into separate melody lines (see the EEG in Figure 2.1 for an example). Thousands of neurons firing together collectively at the same frequency is what makes detecting the waves through the skull by EEG possible. Troughs indicate neurons at rest or inhibited. Frequency is important, but the height of the wave, or the "amplitude," also plays a role in the patterns, indicating strength.

We can use measures of these patterns to map all kinds of processes to their associated locations in the brain. And as you'll see throughout the book, we can take advantage of what we learn to try to manipulate (or entrain) the waves from outside as therapy. In exigent cases, we can do it from the inside with the hope of obtaining desirable outcomes, such as an easing of depression.

What do different types of waves tell us?

Delta waves conjure associations with deep sleep,* but they also are linked to our ability to form new communication connections in the cortex and in processing.[9]

Theta oscillations arise from two regions: the cortex and the hippocampus, although the hippocampal rhythms are probably part of a loop that includes the cortex. Look at this series of numbers and try to remember it in order: 594320619. When you did this exercise, theta oscillations probably became more active at your temples (in the temporal lobes) and at the front of your brain (in the frontal lobe).[10] Did you feel it?† In contrast, if you're sleepy, the theta waves likely will show low activity. High theta activity in someone who's resting is considered to signal a potential problem.[11]

Alpha waves, oscillating in the back of the brain, in the occipital cortex, are active in people who are resting with their eyes closed but are not asleep. They also may serve an inhibitory purpose. They can show high activity during attention tasks, but in the sensory areas that aren't required for the task. In areas that are associated with the task, their activity is low. When

---

* I have a white-noise app that I use for sleeping in strange or noisy places, and one of the options is "delta waves."

† Probably not.

you try to memorize something, for example, your success may depend in part on your alpha waves bombarding unnecessary regions, shushing them up so you can focus.

Some research suggests that alpha waves (and maybe thetas too) coordinate the oscillatory activity of the brain.[12] Despite being the longest-studied, alpha waves and their possible subtypes, effects, locations, and roles remain the most mysterious of the waves.

Although beta waves cycle faster than delta, theta, and alpha waves, they don't reach the peaks that the other types do; they stay flatter as they oscillate. They are active when we're awake, and they're associated with perception and motor performance.

The high-frequency gamma oscillation, split into high and low ranges, is linked to all kinds of information processing, including memory formation. It uses a lot of oxygen and energy and is tweaky in the face of low energy availability. It's responsible for how much attention you give to a stimulus. Indeed, it may interact with alpha oscillations in a sort of back-and-forth that helps you register a deer at the forest's edge even as your attention is focused on the road. Alpha oscillations may inhibit attention to what's off to the side as you drive, so you can keep your eyes on the road. But periodically, gamma waves can break through and give you a window to pick up on what's salient, if anything, outside that field of focus, like a twitchy deer.

Oscillations are of immense interest across many areas of "brain improvement." A broad category of approaches relying on our knowledge of them includes tools that link brains and computers or even two brains. In fact, one of the first brain-computer interfaces (BCIs) used alpha waves to control a robot.[13]

DO-IT-YOURSELF MIND CONTROL

BCIs come in several flavors. We can try to use what we know about oscillations to control them ourselves (neurofeedback), use implants for that purpose, or use EEG signal transfer over broadband to facilitate communication between two brains.* Below is an overview of the basics of each of

---

* First done in 2014.

these applications, with an emphasis on those that are in common use for tailoring the brain.

I won't spend a lot of time on invasive versions of BCI until the final chapter. They require brain implantation of devices that can transmit signals and stimulate the brain. They're used for specific conditions such as intractable seizures, which goes well beyond generally accessible nips and tucks to retrain oscillations or silence a barking DMN. One thing to keep in mind is that given the relative newness of most of these interventions, we do not have much information about their long-term effects, wanted or not.

In neurofeedback, the person learns how to control their own neural activity using real-time information and feedback about oscillations, as measured using EEG or MEG.[14] The target region of the brain is selected for the feature being tailored, such as regulation of attention, anxiety, or emotion. After the baseline neural activity of the region is recorded and understood, the person can try various conscious techniques to alter the patterns, such as the ratio of theta to beta oscillations, to something more salutary.[15]

The idea is that, with EEG or MEG feedback, the user will learn how to regulate these patterns and make a real-time connection between specific tactics and achieving the desirable change.[16] Real-time feedback allows a direct association between a strategy and its effects on the patterns. The cues that let users know whether their strategy to modify their brain waves is working can be a lot of things, including visual, auditory, and electrical. Eventually, the expectation is that the person can achieve these changes on their own, without EEG feedback.

SKULL CROSSINGS

Although EEG and MEG obviously provide access to what's inside your head without cracking it open, they aren't typically used to directly target the contents of your cranium. BCI tools that do get under the skull are called transcranial interventions, and they fall into a couple of broad categories.

The first is transcranial direct current stimulation, or tDCS.[17] The user slaps on a couple of electrodes. One goes where it will affect the target area, and the other is applied somewhere else, usually not on the head. If the electrode operates as a cathode, it will apply a weak current to the target area and dampen activity there. If the electrode is an anode, it will apply a weak current and create excitation.

For about fifty bucks, you can order a kit to do this yourself at home,* which is one reason that Alvaro Pascual-Leone cautioned against trying these technologies without appropriate expertise at hand. The direct-to-consumer path for tDCS has, as one researcher wrote, gained attention from the "media with mixed feelings of excitement and concern."[18]

Put me on the "concerned" side of the spectrum, especially given comments from the fourteen thousand members of a single subreddit devoted to the practice. People complain of burns, increased anxiety, loss of bilingualism (!), and other problems.

Some argue that because the current does, in fact, enter the brain, it is definitionally invasive. They assert that tDCS and similarly intrusive tools should be called "neuron-altering" rather than "noninvasive." At any rate, tDCS carries risks that include putting those electrodes in the wrong place, stimulating an area of the brain way off target, and disrupting perfectly healthy excitation/inhibition oscillations of the neurons making up the neural networks. There has even been some hand-wringing about addiction.[19]

You might have predicted that if a direct current (DC) version of transcranial stimulation exists, an alternating current (AC) version must, too, so congratulations on your foresight! In transcranial alternating current stimulation, or tACS, instead of applying a direct current to the target, the cathode and anode switch back and forth, causing the current to change direction, almost like an oscillation.

The expectation with these stimulations is that the current will strengthen associations and connections between the targeted neurons so that they act together faster. Hypothetically, this, in turn, would help you retain what you're trying to learn. The technique is also used to treat conditions such as depression and anxiety. As with seemingly everything neuroscience, no one can quite nail down how well these expectations are being met.[20]

---

* This is not something I'd do.

Where electricity goes, magnetism can't be far behind,* and these non-invasive intracranial interventions are no exception. Like the other forms of transcranial stimulation, transcranial magnetic stimulation, or TMS, targets the alpha, theta, and gamma waves for modification, but it uses a magnet (obviously).[21] A device that pulses with electricity is connected to a magnetic coil that is placed on the head, precisely over the target area. When the electricity is directed through the coil, it produces a magnetic field that can penetrate a couple of inches into the brain. As is the relationship between electricity and magnetism,† this magnetic field in turn produces current under the cranium.[22]

TMS has been in use since the 1980s. These days, the version most often applied in clinical situations is repetitive TMS, or rTMS, which is approved as a treatment for depression that hasn't responded to several medications and for obsessive-compulsive disorder. In contrast, tDCS is not FDA approved for anything.

You might be wondering if all this zapping is painful. Jens Foell has had some experience with transcranial stimulation, having been plucked from the audience during a seminar to participate in a demonstration of the technology. The researcher administering the TMS pinged Foell right over his motor cortex, where it takes in messages from the thumb. It was the equivalent of receiving notification to move the thumb, and when the region was stimulated, Foell's thumb moved in response, all on its own. "I wasn't even aware he'd started," said Foell, who described looking down at his twitching thumb that he hadn't told to twitch. The researcher, he noted, then intoned to the assembled high-school students that "we are all just machines."‡

These approaches tend to produce individualistic responses and have yet to be refined to narrow targets.[23] In addition to persistent concerns about long-term or placebo effects, along with the do-it-yourself ethos that tDCS evokes, ethicists have concerns about the direction of BCIs in general. These include worries about mind control (known as "brainjacking"), especially

---

* A little physics joke.

† In brief, when a current passes through a wire, it produces a magnetic field, and a shifting magnetic field can draw a current.

‡ We are not.

with implantable devices; misuse of the technology to, for example, "improve cognition" in children who cannot consent; large individual variability in outcome and side effects; equity, such as differential access to interventions that truly offer benefit; and even existential questions about how far alterations go before they cross the threshold of what defines a person's identity.[24] We will look at these apprehensions a bit more in the final chapter.

IS THIS A GOOD STUDY? A QUICK PRIMER

Not all studies are created equal. Factors that affect their rigor include the numbers of participants, how much researchers can control for off-intervention effects, and whether they planned the study before they collected the data or are retrospectively analyzing data that already exist.

Studies that simply look at, or observe, data, whether retrospectively (looking back) or prospectively (planned ahead of time), are called observational studies. Statistical techniques allow researchers to control for known factors that might influence the comparisons they're making, but they can't control for everything. Such studies, therefore, always stand a chance that some unrecognized or unmeasured factor affected the results instead of what's being measured. If the factor is common to both groups, then its influence would flatten real differences between them. If it is more common in one group than another, it could exaggerate their differences and make what's being measured look more important than it is. Either way, the factor has confounded the findings. As an example, studies have found that people who drink coffee have an increased risk of lung cancer. That's just a mathematical association. A closer look at such studies has shown that coffee drinkers, at least when this research was done, also tended to be smokers, and smoking is obviously a well-known cause of lung cancer. If an analysis doesn't control for smoking, and if coffee drinkers smoke more than abstainers do, then a lot of things linked to smoking will erroneously look they're tied to coffee consumption too. Observational studies are always subject to this confounding, so the results should be taken as offering hypotheses.

A subset of the observational study is the case-control study, in which cases (people with a given feature or condition) are matched with controls (people who are similar to them in age, sex, and other ways but do not have the feature or condition in question). The analyses in this kind of study will produce ratios of controls to cases. These ratios are the odds that someone with the feature or condition will have certain outcomes compared to someone without it. Let's say that the question is whether or not aspirin use is associated with chronic kidney disease, but you want to rule out a role for sex or older age. Researchers can compare aspirin intake among people with chronic kidney disease to aspirin intake among people who are a lot like them (similar age, same sex, and so on) who don't have the disease. If they were to find more aspirin use in the group with chronic kidney disease, that result could imply that aspirin is linked to increased risk for the condition. It does not, however, mean that aspirin use *causes* the condition. The association is mathematical only.

That's because observational studies show only mathematical relationships: how much outcomes track mathematically with being in one group versus another. They don't reveal anything about whether or not the differences between the groups *cause* the differences in outcomes. Case-control studies may focus on a candidate cause, such as taking aspirin and risk for chronic kidney disease, but they don't *demonstrate* the cause.

Studies that are randomized and controlled address some of these problems. The more a study is controlled, the lower the risk that confounding factors will skew the results. Clinical trials have very specific inclusion rules so that the groups in the trials are as similar as possible. Participants are randomly allocated to those groups using accepted randomization techniques. And they usually use a placebo control for one group and an intervention for the other, and conceal which group is receiving which (called "blinding"). That way, among a lot of similar people, the only difference is the exposure to the intervention or the placebo, and no one knows who's exposed to what.

One feature that can weaken any type of study is including only small numbers of participants. That matters whether the studies are observational or another kind. The trials for the COVID-19 vaccines enrolled tens of thousands of people, allocated randomly to receive the candidate vaccine or a placebo injection. The results of these trials are quite robust.* If the

---

* Even more so because by including this many people, the researchers also could analyze various subgroups by age, health conditions, and ancestry/ethnicity.

researchers had included only a few dozen people in each group, the findings would be extremely weak and not very useful. When a study involves few people, just a handful of participants with extreme responses could easily skew findings for their entire group.

The most frustrating thing I encounter in neuroscience research[*] is how rarely anyone seems to replicate other studies to see if they, too, get the same results.[†] In fact, a defining feature of so much of the field is that no one uses the same terms, definitions, outcomes, participants, intervention parameters, or study design. That means when you see a result and want to find out if someone else has confirmed it, you can't.

One solution for this problem—and the problem presented by having a lot of tiny, different studies ostensibly of the same thing—is to perform a meta-analysis. Meta-analyses rely on a methodical approach to finding all the studies associated with a specific research question, such as "Is aspirin associated with reduced rates of colon cancer?" Authors can limit the types of studies they include to observational or case-control or randomized trials, but still the search and inclusion steps follow certain guidelines. Researchers then pull the data from these studies, combine them, and perform specific analyses on the whole shebang. They also will grade the evidence on a scale from very low certainty to high certainty. In the absence of replications and studies with tons of participants, this is the best way to sort out signal from noise.

CHECKLIST: TELLING REAL FROM FAKE SCIENCE

We live in a time when news of real events is labeled fake and people can digitally doctor videos to create "real" events that never happened. It has become more important than ever to develop a critical eye. That's always been the case when it comes to assessing claims made by people trying to sell you something as "scientific."[‡]

---

* And a lot of other areas of research too.

† Things may be taking a turn on this front, and a few journals focus entirely on publishing replications of studies.

‡ This section is based on an article I wrote for the *Forbes* contributor network.

There are plenty of examples of pseudoscientific claims that have raked in the bucks, thanks to a convincing assertion that the science backed them up. Developing a nose for real evidence versus nonsense dressed as evidence is crucial if you're considering doing anything to your body, and, I'd say, especially to your brain.

But discerning real from fake science can be tough. Clever purveyors can use just enough sciencey-sounding language and a little bag of truthiness tricks to come across as plausibly valid. So how can you figure out that's happening?

Here's your checklist for Chapter 2: ten questions you should always ask and answer as you consider interventions for your brain.

1. *What is the source?* Dig into who is running the "advocacy" website or fansite for a product or intervention. Find out whose money is behind it.

2. *What is the agenda?* Conflicts of interest matter. Check the funding sources and affiliations in a published study or on a website selling the product.

3. *What kind of language does the promotional info use?* Are there lots of emotion words or exclamation points along with language that sounds highly technical or jargony? If you're not sure, select a term and google it, or ask a scientist if you can find one, perhaps via social media.

4. *Does the product involve testimonials?* If testimonials are the only supporting information on offer, tread very carefully. Anyone can write a testimonial and publish it. Here's an example I once used to make this point:[*]

> I felt that I knew nothing about science until I found this site! Now, my brain is packed with science facts, and I'm earning my PhD in aerospace engineering this year! If it could do it for me, the Science Consumer site can do it for you too! THANKS, SCIENCE CONSUMER SITE!
>
> —xoxo, Julie C., North Carolina

---

[*] I made this up specifically for this purpose for a blog I once maintained.

5. *Are there claims of exclusivity?* Typically, new findings arise out of existing knowledge and involve the contributions of many, many people. It's quite rare that a novel therapy or intervention would spring up overnight or that one giant of industry would discover it. Also, watch for words like "proprietary" and "secret." These terms signal that the intervention has likely not been exposed to the light of scientific critique.

6. *Is there mention of a conspiracy of any kind?* Claims such as "Doctors don't want you to know" are extremely dubious. In addition to the fact that "doctors" aren't a monolithic entity, why wouldn't the millions of physicians in the world want you to know about something that might improve your health?

7. *Does the claim involve multiple unassociated conditions?* Promises that a specific intervention will address everything about your brain and make it perfect are nonsense because of the complexities underlying different conditions.

8. *Is there a personal reputation at stake?* Sometimes, money isn't the only thing that motivates people. Some people just don't want to be wrong, or they're addicted to their role as a contrarian or "savior" and don't want to be viewed as fallible.

9. *Were real scientific processes involved?* Evidence-based interventions generally go through many steps of a scientific process before they enter common use. These steps might include basic research in cells and animals, clinical research with patients/volunteers in several heavily regulated phases, peer review at each step of the way, and a trail of published research papers for each study, building the evidence base.

10. *Is there expertise that's relevant?* During the COVID-19 pandemic, people who had zero training in epidemiology, virology, or infectious disease gained huge social media platforms by mouthing off plausibly, all for the sake of self-aggrandizement. Closely check the subject-matter knowledge of anyone trying to sell you an intervention. I also recommend listening to more than one expert.[*]

---

[*] So please go ask other science experts about this list.

CHAPTER 3

GLOBAL COGNITION I: WHY WE'RE DOING IT WRONG

W HAT IS COGNITION? If you just thought about that question, then you were using cognition. It is thinking—a process that takes sensory inputs and existing knowledge and applies them along with native insight to solve problems old and new, to use language, to form and store memories, and to use judgment in all of the above. That's a lot. And it doesn't even mention the intrusion (or inclusion, depending on your attitude) of emotions, which paints these activities in the varying hues of mood and feeling.

Intelligence is something a little different. The concept of intelligence is supposed to capture the contours of our cognition, giving it shape and magnitude. Intelligence is sometimes divided into broad categories. One is a crystallized form, built from nuggets of knowledge we take up and retrieve as needed, some of us doing so easily and capaciously and others less so. The other is fluid intelligence, the flow of ideas and understanding through the byways of our minds, showing us how things are related (sometimes in newly discovered ways), grasping the abstract (the distillation of the

DMN as shorthand for "self" is an example of abstracting), and solving new problems.

These terms get mixed into a word salad that conflates the various ways we talk about our thinking skills. One of those conflations is to use "IQ," or intelligence quotient,* as a proxy for intelligence. It's not. It's really a statistical estimate of some portion of intelligence, an observable end point derived from testing, a tool for those whose area of study is psychological measurement. They seek to develop tests and measurement tools and other doodads to try to capture the complex landscape of the human brain.

The IQ value, derived from such tests, has been associated with something theoretical and still quite fluid, which is our cognitive ability, or g. As you will see, tailoring this g will probably have to wait given that we have yet to fully grasp its true contours. But that has not stopped people from using it as a metric for studying (and at times selling) various cognitive interventions and trying to make racist arguments based on it.

Studies highlight a role for the DMN and other networks in cognition, including studies of people with neurological conditions such as multiple sclerosis[1] and stroke, which can be associated with compromised cognition. Researchers can't ethically shut down a network on purpose in humans to see what happens. But the regions of damage caused by these conditions and their symptoms can sometimes open the door to linking a physical area of damage with the traits that damage affects. Specifically, studies looking at damage to a deep brain structure called the thalamus, which relays signals to and from the cortex, and at disrupted DMNs highlight important roles for both in cognition.

These two are not alone, though. The other circuits I highlighted in Chapter 1 are also involved, including the CEN and the salience network.[2] Their ability to rev or relax together or in opposing directions underlies our ability to engage in complex cognition.

In stroke, cognitive impairment is common. In a study that assessed the three primary networks believed to be involved in cognition—DMN, CEN, and SN—researchers found that people with stroke have less connectivity

---

* So-called because it originally was the result of a division problem: the person's mental age score divided by their actual age and multiplied by one hundred.

between brain hemispheres than people with no history of stroke. The authors of that small study examined the network connectivity on the side of the stroke, on the opposite side, and between hemispheres. People had better cognitive performance at six months poststroke if their baseline just after the stroke showed a more intact DMN on the stroke side and better interhemispheric connections.[3]

Cognitive ability also relies on some flexibility among the networks and their hubs. Intriguing research suggests that depending on the task we're completing, some nodes can work with different networks, rather than showing fealty to a single routine.[4]

In keeping with our Planet Brain metaphor, you can think of it this way: If we want to send a package from New York to San Francisco, there is a series of hubs in between that we may or may not include in the process, depending on logistics. Perhaps we send the package straight through, leaping right over Chicago. Or perhaps the package needs to be added to a bundle of other packages in Chicago for our purposes. In that case, we make a stop at the Chicago hub before we move on to San Francisco and complete the process. Chicago didn't disappear when we leapfrogged over it. It just wasn't used in a specific scenario. These various reconfigurations aid the communication network in achieving different but globally related aims. The good news is that this reconfiguration flexibility likely means flexibility in our cognition.

Other studies highlight a role for the DMN in a suite of cognitive functions, including memory, attention, and social behavior, and implicate its derailment in conditions in which cognition can be compromised, including schizophrenia.[5] That may be in part because the DMN, true to its reputation, is a self-referencing attention hog, so everyone's paying the most attention to it. As we've seen, other networks, obviously, are involved too.

When it comes to the CEN and DMN, age matters. There is an increasing tendency for these networks to operate in anticorrelation mode as a person grows, going from apparently chaotic correlations in early childhood, to a more predictably mixed association in adolescence, to anticorrelation into adulthood. Children who show higher anticorrelation between the two networks score higher on IQ tests, suggesting that the maturation of this relationship is important to what such tests capture.[6]

CAN YOUR IQ CHANGE?

In 2013, journalist Lauren Cox asked five experts the question "Can your IQ change?" The answers?

> Depends.
> Yes.
> Absolutely.
> There's no such thing as an IQ.
> Depends.

They weren't talking about changes wrought by way of "brain-training" exercises and such but about changes over the lifespan and at the population level.

The experts emphasized that teaching a skill, such as planning, can improve function more globally, potentially translating into higher IQ scores. They talked about how IQ scores are most volatile in childhood through adolescence* but then generally settle into a fairly stable narrow range as we get older. They questioned the validity of the score as a reflection of inborn capacity.

And the expert who dismissed IQ as not a real thing, Alan S. Kaufman, clinical professor of psychology at the Yale University School of Medicine, remarked that an IQ value isn't "like stepping on a scale to determine how much you weigh." For someone with a score at a given point in time of, say, 126, he said, the real range could be anywhere from 120 to 132. Take a different test, and the score could be different.

These scores aren't the metrics we're used to in our day-to-day lives. We measure distance in miles or meters, and mass in grams. The mass in kilograms of a gold brick will be the same on Earth as it is on the moon, but IQ varies based on context. It's compared in a population of the same age and sex. The score a person is assigned depends on how they compare to that population by age. And the scale is not geometric or exponential or absolute in any way.

---

* One of the experts called the fluctuations, which can be more than 20 points, "enormous."

As one researcher put it, someone who scores an IQ of 130 is not 30 percent smarter than someone who scores 100. The person who scores a 100 is right at the center of the scores for that population for their age, at the median, the 50th percentile. A person who scores 130 will land a couple of increments away from the center, in the 2 percent end of the population distribution, around the 1.4th percentile. Bumping a score from 100 to 103 kicks the test taker to about the 58th percentile, but an increase from 133 to 136 means edging up by only 0.5 percentile.[7]

Those peddling interventions might administer a test before and after whatever remedy they're hawking and then report that the test takers' average score increased, implying that the group's intelligence did too. Yet the increments aren't equal, and that increase—if true—doesn't mean an equal magnitude of effect for each individual.

A FRAUGHT HISTORY

The history of IQ is largely a racist and classist one. Thomas J. Bouchard Jr., professor emeritus of psychology at the University of Minnesota, wrote in 2014 that "many species have evolved a general-purpose mechanism (a general biological intelligence) for dealing with the environments in which they evolved." The statement reverses the relationship of trait to environment—species don't "evolve mechanisms to deal with an environment." Nature selects for traits (and underlying gene variants) that give individuals an edge in a given environment. What the statement does imply, at any rate, is that through the course of evolution, environment shapes this "general biological intelligence."

That points to two obvious takeaways: Humans evolved in Africa, and whatever that environment was, it's what shaped us as a species. We've all been humans ever since, and we've traveled all over the world in the ensuing fifty thousand to one hundred thousand years or so, doing much to create our own environments as we've gone along. But those aren't the environments in which we originally *evolved* as the species we all are. Most of our shared evolutionary history is in the same environment—an environment that included each other as strong selection pressures—so generally speaking, we all have similar adaptations to that experience.

The evolutionary past common to all of us hasn't stopped people from making desperate efforts to couple intelligence and the archaic concept of biological "races" of humans. The same people interested in the eugenicist goals of stratifying by "race" have spent decades using tests to justify their beliefs and fulfill those aspirations. Right along with them have been other people who have pointed out, over and over, example after example of the bias in intelligence tests and the effects that factors such as socioeconomic status and even birth weight can have on outcomes.

After the first IQ tests emerged at the turn of the twentieth century, there were immediate hints that environment could be a decisive factor in outcomes. Reports from as early as 1922 indicated that concerted efforts to educate children could lead to substantial increases in scores.[8] In one example, a child entered a training program at age seven, unable to read. Her mother was a widow who was "very ignorant," and the family's overall standard of living had "always been low." The child scored 95 on the IQ test administered to her* and "had very little language, although she is American born." After two years of school, she was tested again and scored 111, shifting her from below to above average in IQ.

A 1929 study of "imbeciles," "morons," and "idiots"† who were tested at the beginning and end of a study period found that at least one in ten showed an increase in scores.[9] The study author noted that the changes could shift "high-grade defectives" out of the running for being institutionalized and leave them to the mercies of independence, something the author deplored. The influence of education will come up again as the one intervention that affects IQ, and social inputs may have something to do with that.

IQ was an obsession for a certain demographic from its inception, used to confirm ideas about "classes" of people deemed "lesser" or "depraved" in some way, including unwed mothers, Black people, Indigenous peoples (with some publications even reporting IQ scores by tribal identity), people with scabies, and "delinquents." The main utility of the tests seemed to be to

---

* The test at the time was Binet.

† All considered clinical classifications at the time but certainly leveraged to dump people into categorical buckets and leave them there to rot, including in institutions, which is why I do not use these ableist terms outside quotation marks.

confirm a preexisting negative bias and produce evidence to defend keeping certain types of people in their places.

Doing this wasn't viewed as deplorable. One luminary of this crowd was Raymond Cattell, a psychologist from the twentieth century who was also known as the "father of trait measurement."[10] The apparent reason he fathered this field is because he wanted to make sure undesirable types wouldn't be fathering anyone themselves. He was considered an eminent psychologist, widely published and highly respected, at least by some.

For these reasons, in 1997 the American Psychological Association (APA) decided to award Cattell, age ninety-two at the time, with its Gold Medal Award for Life Achievement in Psychological Science. The only problem was that his lifetime of notable achievements, many of them related to measuring the "trait" of intelligence, included making some remarkably bigoted assertions. Association members protested the decision, and ultimately the board elected to postpone the award because his "lifelong commitment to eugenics and racism . . . will be a moral stain on our discipline and our organization." Cattell ultimately withdrew himself from the nomination, but not before newspapers took up the banner and dubbed him a "Victim of Liberalism." Some of Cattell's supporters and colleagues resigned from the association in protest of the board's decision.

These brave people were choosing quite the hill to die on, given that Cattell was known for writing such gems as "A race such as the Negro had to be brought to euthanasia not by war or brutality but by gradual restriction of births and by life in adapted reserves and asylums." He coined the grotesque term "genthanasia" to describe his proposal to "phase out a race . . . by educational and birth control measures without a single member dying before his time." He was no less bearish about Jewish people, for their practice of "living in other nations."

Despite the record, in his letter withdrawing from the APA nomination, Cattell claimed that critics had "twisted" his position. Yet as late as 1987, he was proposing brilliant philosophies such as "beyondism," which was intended to focus on strictly segregating the "races," determining which ones were the best fit to be retained and systematically eliminating the others. If races were allowed to mix it up, he argued, the result would be a "parasitism," with the members of less worthy races siphoning the strength of the superior (of course, White) race.

Why all this about Cattell? Because he was hugely influential—as his citation numbers attest—and especially so as a backer of the value of intelligence testing to sort out who deserved to be genthanasia'd or not.[11]

SPOTTING THE G

Cattell got his start as a grad student studying under Charles Spearman, a psychologist at University College London. Spearman and others had noted that people who do well on one test of intelligence seem to do well on others, regardless of the aspect of intelligence being measured. Some factor, Spearman reasoned, must be contributing to this common performance or ability. He dubbed that unseen factor g, or Spearman's g.

Although his original single-factor proposal was quickly falsified, the core idea persists that a mysterious g drives the correlation among the scores, but the factor itself was and still is a hypothetical. It is subjective. That means the researcher hypothesizing what it might be is injecting bias into it. And if you decide that the factor is something inherent or inborn and accordingly create tests to assess for the factor, that seems like getting trapped in a tautology.

The correlations across such tests are real.[12] But the factor that unites them—the "common fundamental function" that contributes to good performance across the board, regardless of what the specific tests address—is the elusive desideratum that cognitive scientists and brain trainers have been chasing in the 117 years since Spearman introduced it.*

Other candidates for the factor exist. Parental support, physical and mental well-being, stress, and motivation can be universally influential on cognitive test performance. Motivation emerges as a powerful influence. In one study, after researchers adjusted results for the effects of motivation for taking the test, the association between scores and success in life disappeared. The implication was that the unifying factor in the associations was merely a real enthusiasm about test taking.[13]

---

* Spearman himself reportedly did not believe that g could be captured by a single test.

Reinforcing this idea, other researchers have found that paying people to take the tests gets a better performance out of them, to the tune of more than half a standard deviation.* Such findings suggest that the tests may measure intelligence, sure, but they also capture the level of motivation for taking them.[14] Perhaps the true way to increase scores on intelligence tests, whatever the scores mean, is to pay the test takers.

Once people began to believe that the *g* was intelligence, leveraging it to exclude certain groups of people and to bolster eugenicist arguments was not far behind. These efforts have shadowed both IQ and arguments about *g* and its alleged genetic origins ever since. Cattell, writing in the *Eugenics Review* in 1936, was deeply concerned about declining national intelligence and people having too many children.[15] His worry was that civilization would decline as environmental factors "sterilize and remove" the more intelligent "strains" of humans. Necessarily, genthanasia had to follow.

To bolster his argument that "less intelligent strains" of humans were having too many children, he conducted a study of ten-year-olds living in an English city (with a whopping 239,000 people) and of children from "one unspoilt rural area" of towns with populations ranging from 300 to 6,000. Finding lower scores among the rural children, he concluded that "all considerations indicate that the urban superiority is due to a great readiness of the more intelligent families to migrate to the cities. . . . The more gifted biological strains will be in time found where the more attractive conditions exist." Or perhaps being around a lot of different kinds of people is a factor in expanding our minds—not only figuratively, but in terms of sprouting new physical connections in our brains as we regularly make new discoveries about other human beings while learning their stories.

Cattell argued that people who scored low on his test were having more children than people who scored high, resulting in a surplus of children from "less intelligent" stock. Sure of his arguments, he finished up with a list of claims intended to support an almost total genetic basis of intelligence.

Toward the end, pausing to wring his hands, one presumes, he addressed the divine muse of cognition: "If the able group is cut down in this way,

---

* Which means larger effects for those with lower scores to begin with.

whence are the skilled doctors and surgeons, the organizers and politicians, the writers and research workers of the next generation to come?" It was a straight line from these assertions to his writing, as late as 1987, about the tragedy of "racial and cultural slumping," or treating people of different racial backgrounds the same way. And he had specific political plans: he felt that Black people should be deprived of the vote in the United States by way of intelligence testing that would exclude them.

The role of scientists in Cattell's view—indeed, the motivation for the father of trait measurement's interest in trait measurement—was to provide the data necessary to determine which "racio-cultural" groups were best suited for evolutionary advance and which should be left behind. Such "scientific" judgments were then to be translated into action: "successful groups" were to expand and increase their power and influence, while "failing groups should . . . be allowed to go to the wall."

Nor would Cattell's system allow a "failing group" to enjoy any charitable assistance from others. In his view, that would only reinforce the strength of the faulty culture and postpone the reduction of genetic defect. To ensure the appropriate result in such cases, Cattell encouraged his idea of genthanasia—a process of "phasing out" a "moribund" group, not by violent means but through "educational and birth control measures." William H. Tucker, who wrote about Cattell in his 2009 book, *The Cattell Controversy: Race, Science, and Ideology*, told an interviewer that Cattell sired "trait measurement" specifically to produce data for identifying the "strains" that should be targeted for deletion from the human population—or genthanasia.[16]

Whenever people talk lovingly about IQ or want to use it to define a human being, including themselves, I think about its birth and why it was begotten in the first place. And then I think about how many Cattells are still out there, flogging IQ scores and *g* as a foundation for establishing a race-based intelligence hierarchy that, combined with other race-based rationalizations, leaves them at the top.

THE MIGHTY FLYNN EFFECT

Intelligence tests vary in how much *g* they capture, but the scores are all supposed to be "normed" to a representative "normative" population and the score distribution for that population. Given the changeability of scores, these populations themselves must be updated, or they will no longer be normative.

An example of why new norming is needed is the "Flynn effect," also known as "secular IQ gains" or the "secular rise in IQ scores." James Flynn, an intelligence researcher at the University of Otago in New Zealand, documented the phenomenon that in modern societies, IQ scores have been rising generation over generation, Cattell's 1930s agonizing about declines notwithstanding.[17] "If you scored people 100 years ago against our norms," Flynn said, "they would score a 70. . . . If you scored us against their norms, we would score 130," or borderline gifted.[18] It's not possible for changes at the genetic level to explain this rapid and global upward shift.

IQ is a psychometric value grown from eugenics. It is kept fed and watered by hereditarians who seek to use every new genetic technique to shore up its putative validity as a reflection of native ability. Yet somehow, despite their urgent claims that intelligence is inborn and that IQ reflects this inborn capacity, in industrialized societies, people have been scoring increasingly higher on IQ tests since the mid-twentieth century. Within the span of fifty years, the average score has increased by 18 points. It is not possible, given the inferred contribution of hundreds of gene variants to our neuronal power, for this value to reflect something evolutionary within the span of a generation. Even if intelligence were a matter of a single powerful gene variant, an increase of that magnitude in its frequency wouldn't manifest across geographically distanced modern industrial societies within five decades.

Thus, something else that's not genetic must be the g factor in this case. One idea, known as the mutualism model, posits that all our cognitive functions interact in their mutual development with each other. This interacting ecosystem of developing cognitive functions would in turn explain why test scores capturing one function correlate with scores capturing another.[19]

THE TESTS

The tests themselves have changed over time, although that fact has been set aside as an explanation for the Flynn effect. Alfred Binet developed the initial versions, and his name is still on the Stanford-Binet Intelligence Scale in use today. Other tests include the Wechsler Intelligence Scale for Children, and the Wechsler Intelligence Scale for Adults. These tests usually don't only return a global score in the context of the normed testing population but also give scores in subcategories such as verbal and nonverbal reasoning.

Studies suggest that as people take tests purported to measure intelligence, the hubs and their networks in the brain that show the most inferred activity are, as you might expect, associated with executive function and attention. This activity correlates with global scores on the tests.[20] Given that these hubs and their networks seem to start out a bit chaotic in childhood before settling into anticorrelation between the DMN and the CEN through adolescence, perhaps it's not surprising that scores during these periods can fluctuate. At young adulthood, they tend to stabilize until older age.

As a way to cull out cultural effects, the tests focus on reasoning that doesn't require great verbal facility. An example is shown in Figure 3.1.

Just what do the tests measure? Lazar Stankov, an emeritus professor of psychology at the University of Sydney, who has a decades-long history of research in this field, offered me his take on it: "They are measuring individual

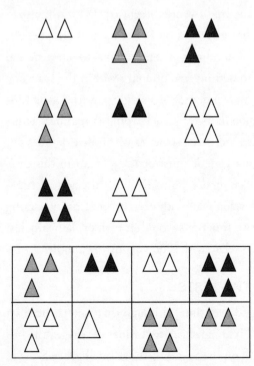

Figure 3.1. What is the correct choice for the next pattern in the sequence? Solution on page 75.* From the Sandia Matrices, a free alternative to Raven's Progressive Matrices, after Harris et al. 2020.

differences in people's ability to solve cognitive tasks . . . your ability to look at things from a different point of view. So they are basically telling us how good a person is at solving problems."

He said that when people add in considerations of sensory processing, which has "a cognitive side" but was something that Spearman and his acolytes dismissed, the idea of IQ and the g becomes less important. That sounds similar to how the notion was diminished when researchers controlled for motivation. And it may be one explanation for why education and especially an important classroom intervention in Belgrade in the mid-twentieth century may have yielded real effects on IQ: the human touch.

IS *G* IN THE GENES?

How important is g, and how much of g can be explained by variations in our genes? The answers to that question are all over the map. It depends on the kind of study being performed and which intelligence measure is involved. A recent report based on data from the UK Biobank found that a "genetic g factor" explained about 58 percent of the genetic variance in traits,* but across the seven traits measured, the genetic contribution ranged from 9 percent to 95 percent.[21]

Twin studies get a lot of attention because identical twins are presumed to have identical genetic complements. That can change over time, though, thanks to a process that effectively hides parts of the DNA, leaving a genetic sequence unchanged but making it unavailable. This process leads to the accumulation of differences even between identical twins in how the genes are used, beginning before birth. Scores can be similar between identical twins, but the twins that are heavier at birth than their identical sibling tend to score higher early on; even more interesting, children who aren't twins tend to score higher than children who are.[22] These findings certainly suggest an influence, starting in utero, of factors unrelated to the genetic variants an individual carries.

Note that when people talk about heritability in these studies, the math isn't as straightforward as you might think. If someone says the heritability

---

* One expert told me that the method the authors used to calculate heritability is known for overestimating these values, by as much as fourfold in some cases.

is 0.5, that doesn't mean the genes invariably contribute 50 percent to that trait and the environment does the rest. Heritability is a lot more compli-cated than that and doesn't describe the status of any one individual. Briefly, the concept means that in *that* specific study of *those* people in their environ-ments, genes on average were associated with about half the total effect in shaping the trait being measured.

A study that assesses thousands of tiny changes across the genome and their links to intelligence—called a genome-wide association study, or GWAS—offers another way to eyeball the effect of gene variants. Results of such studies suggest that each little change has a teeny effect on its own. Collected together in a "polygenic score," they still show only small associa-tions with IQ, or explain only a small portion of the variance in IQ scores.[23]

If these collective changes add up to such a small effect, how do we explain the results of twin studies that people point to as showing a high shared genetic contribution? Twin studies don't exclude environment, which twins usually share from conception, whether they are identical or not. In addition, a parent's gene variants, even if they don't pass them to a child, can affect how a child uses their own. This "dynastic effect" is called "genetic nurture." Genetic nurture can begin prenatally, with conditions before birth explaining about a third of what can look like direct genetic effects on a child's outcomes in the early years. That proportion of the influence is only indirectly genetic, though, related to gene variants the birthing parent has, rather than ones the child carries.[24]

One way around the influence of genetic nurture is to compare scores for people who are not related but share genetic similarities, based on thousands and thousands of examined DNA changes. In such studies, the heritability, or effect of the collected gene variants, amounts to between 20 and 30 per-cent of the explanation for cognitive test scores.[25] In yet another approach that looks at the effects of some of the less common variants among related people and controls for environment, the heritability goes back up to 54 percent. Because the methods used in these studies can inflate heritabil-ity, some authors have tried different ways to address this inflation, finding even lower values for heritability and a greater contribution of environment.

Researchers using these adjustments to assess heritability of educational attainment in Iceland found it to be only 17 percent.[26]

These heritability values can also change over time and from group to group, another reason they have been all over the place during the decades of attempts to pin down a number. In the meantime, that IQ, whatever it represents, has been climbing, a manifestation of the Flynn effect, one that is showing up on nonverbal tests. In Great Britain alone, scores climbed by 27.5 points on a test with a maximum score of 60 from 1947 to 2002, Flynn found.[27] The huge change would represent in some cases a shift of the population from "normalcy to widespread giftedness," which no one seems to think is likely.

Flynn hypothesized that the demands of modern living and modern education ingrained more skills in abstract problem-solving, which in turn manifested in better performance on the tests. In other words, a few years of environmental entrainment on solving such problems led to better performance on a test of solving such problems. Learning meant performing better. If you're sensing a theme, you're not imagining it.

DOES IT MATTER?

As one IQ researcher told me, "I have trouble trusting both the central tenets and results of most IQ research. . . . A lot of it has to do with privilege, with societal factors that determine whether or not you're going to be successful in life." They continued, "You take out all of these factors that have nothing to do with innate ability, and there's nothing left after that. I don't think that there's a need to define an ability that defines you as a person." A lot of the more serious IQ researchers, they added, accept IQ as something that is malleable.

Whether or not any of this matters depends on what you mean by "this" and "matters." Bumping your personal score by 3 or whatever points? Not much to get excited about. But. Scores on IQ tests, whatever they are measuring and whatever the g really is, unquestionably show mathematical associations with outcomes later in life. They are related to occupation, health, and overall success as society tends to define it.[28] The way the correlations go, if you're interested in boosting your cognitive power, your best bet is to travel back in time, experience specific factors in the womb (possibly such as being a singleton conception), revisit your parents and their socioeconomic status and make everyone comfortably wealthy, and be sure to have a great deal of motivation to do extremely well when taking the tests.

All of that impossibility aside, is there any benefit to learning an individual's IQ score? Sure. If you know their baseline and there's a dramatic change (probably downward), that could signal a specific condition affecting their thinking. And given that IQ scores, regardless of their underpinnings, show associations with later health and relative wealth, perhaps attending to social factors that seem to increase scores might help lift a lot of boats besides your own. But that doesn't mean those boats had an ancestral predisposition to an IQ lower than yours, even though some researchers continue to try to link the two.

What does link to IQ is years of formal education (theme alert!). For example, one study of the effects of education on IQ found that raising the age at which children are permitted to leave formal education—in this case, from age fifteen to age sixteen in the UK—was associated with increased IQ *and* with certain health and other outcomes.[29] Those factors walk hand in hand. Hence, whatever IQ is measuring, the number can be meaningful as a predictor of outcomes, even as it cannot capture the whole capacity of a human being.

The authors of the UK study took advantage of a 1972 change in the law mandating an increase in the legal age (from fifteen to sixteen) at which one could leave school. They compared results for students who had chosen to stay beyond age fifteen before the law took place (85 percent of them) to those for students who were required to stay at least a year after age fifteen (100 percent).

The data covered 502,644 participants in the UK Biobank. The results showed increases in intelligence scores among people who stayed in school beyond age fifteen.[30] The comparison between children who had a choice and those who had to stay demonstrated that something beyond intrinsic motivation was a factor in the increase. Was it education? Considering that in addition to parental income and birth weight, education has some influence on intelligence, perhaps education is where we should be looking.

How profound is the correlation between IQ score and later outcomes? Estimates of a "typical effect size" suggest that a relative IQ one standard deviation higher at a younger age could translate into a risk reduction as high as 25 percent for later illness and mortality at a given time.[31]

Most interventions that promise to expand your IQ won't reach effects of this magnitude, which should be considered in your cost-benefit analysis of their worth. Giovanni Sala is an assistant professor in the Division of Systems Medical Science at Fujita Health University in Japan and an author on two crucial studies I discuss in the next chapter related to "brain training." He told me, "I never met anyone who was born with an IQ of 100 and ended up having an IQ of 130" as a result of any intervention.

Testing IQ may not capture some ground truth about a person's potential, much less capture their full human capacity, but it does serve as a portent and a portal. The portent is that IQ scores can be predictive of outcomes seemingly unrelated to intelligence, even those we attribute to genetics, such as longevity.

The scores are therefore important indicators, if not in the ways people sometimes seek to abuse them. Their role as a gauge of social outcomes opens up the portal: IQ may not truly reflect neural potential or capacity, or be a good measure of either for lots of reasons. But it sure does capture how fiscally and physically well off someone is and will be during their adult lives, and it does so at a very early age, before the teens. This period offers us a window in time during which we can boost factors that underlie the performance associations, the *g*, across tests.

One factor turns up again and again as compellingly and causally affecting IQ, as you'll see in the next chapter, which features a roundup of the various interventions people claim will bump it up.

---

* The one in the bottom right box.

CHAPTER 4

GLOBAL COGNITION II: GAME TIME?

Tom Brady was chosen in the sixth round of the 2000 NFL draft—the 199th pick—and went on to win seven Super Bowls (backed by great defensive lines).[1] In 2020, six wins into the season, he left his original team, the New England Patriots, to join the Tampa Bay Buccaneers, promptly returning with that team for his ninth Super Bowl and that seventh win. (At age forty-three, he also moved into second place for all-time greatest number of sacks, having succumbed to 521 of them during his longer-than-usual quarterback career.) His stated goal is to play until he reaches age forty-five. He's sooooo close.

Among all those sacks, three of them in a single recent game, he has sustained at least one concussion (and likely more—he is human, after all).[2] He has also spun out a sideline industry of a personal brand, the TB12 Method, which promises that a specified regimen of exercise and nutrition—along with drinking an untoward amount of water each day—will make you sharp and successful, just like Tom. One aspect of his routine, according to reports, is to play dozens of brain games every day. He has said that doing so helps him with attention, memory, and processing speed. These claims receive breathless coverage from sports fans and sports journalists,* but how likely are they to be real?

The games he uses are the work of BrainHQ, which Brady has said he heard about from neuroscientists who were part of his training team (we should all be so lucky—but with all those sacks, availing himself of such expertise seems prudent). In his book describing the TB12 Method, Brady wrote that the exercises enable him to take in more of what's going on around him. More accurately, they facilitate faster and better decision-making. You can see where this skill might be useful for a quarterback, but he was, after all, a good quarterback and winning Super Bowls before he first came across the techniques. Perhaps being a quarterback for a really long time is what powers one's successful and accurate plays, rather than playing computer games for several hours a day.

Curious about this activity and interested in possibly leveraging my performance into becoming a professional quarterback myself, I went to the BrainHQ website, which offers visitors the chance to try out a few of the games. The first one I played involved half a dozen butterflies whose position I had to track in quick flashes on the screen. The flashes presented visually different versions of the butterflies, three of which were identical. The user was to track the locations of the three identical butterflies before they vanished, and then click with the mouse in each location.

The better you perform, the harder the game is supposed to become; conversely, if you start clicking incorrectly, the game supposedly gets easier. I gave it a shot and was almost immediately Bored. To. Death. I tried to imagine a forty-three-year-old, seven-Super-Bowl-winning, millionaire pro athlete hulking over a laptop, tagging the memory of flashing butterflies, diligently, for hours every day, and . . . let's just say I was amused.

The company claims that these exercises target specific aspects of attention, "brain speed," memory, intelligence, and even people skills. The website says that the "cognitive benefits are proven." To back this up, they offer

---

* Sample breathlessness, from CNBC coverage: "Brady's mental strength became most apparent in the second half of last night's game, when he seamlessly drove the ball 69 yards in five plays, setting up his running back teammate Sony Michel's two-yard touchdown in the fourth quarter" (www.cnbc.com/2019/02/04/the-surprising-brain-exercises-tom-brady -uses-to-stay-mentally-sharp.html). That's some multifaceted mental strength.

what they call "100+ published papers." But such stacks of papers can be a classic example of "salami slicing," which we discussed in Chapter 2.

It's certainly not new to the genre. In a meta-analysis of training on action video games, Daniel Simons, professor of psychology at the University of Illinois at Urbana-Champaign, and his colleagues found multiple papers describing different outcomes from the same studies.[3] It's the research equivalent of making your font size twenty points instead of twelve so you can reach the target page count your instructor requires. In slicing salami, a research group performs an intervention in a single study but then reports findings in separate publications, slicing and dicing subgroups or outcomes in different ways. The result is a piled-up publication count that does not represent the number of actual studies performed. But do these programs do what's being claimed? "It depends on what the goal is," Simons told me. If the goal is to be able to track butterflies on a screen and accurately recall and mark their location, that's one thing. But if your goal is enhanced cognition—something that extends beyond a skill associated with tagging butterflies—then disappointment likely awaits?

Brain-training programs promise a lot with only a modest investment of time and money, so it is not surprising that they are popular. Sadly, their potential has not lived up to their hype. But one pattern showing hints of tiny effects does emerge: any improvement tends to be most noticeable with worse baseline measures.

If Tom Brady has sustained concussions, perhaps his baseline after several hundred sacks will not be at the same level as that of someone who has spent their time running marathons instead of being repeatedly tackled. Maybe, in some narrow zone of skill, he is experiencing improvement. Who knows? What does seem clear after a deep dive and a long swim around in the evidence, and after chatting with people who dive even deeper, is this: only two approaches seem to enhance global cognition in any way, and neither of them is a game or a pill.*

In the meantime, if you want to emulate Tom Brady and spend your time tagging butterflies on a screen, BrainHQ makes this opportunity available

---

* An early cliffhanger for the next chapter.

to you for a monthly subscription of fourteen dollars (or a yearly one of ninety-six), with a thirty-day money-back guarantee.

CAN WE TAILOR IT LIKE TOM DID(N'T)?

Research suggests that, sure, the DMN, CEN, and their connectivity show associations with how people perform on intelligence tests.[4] But they also show that this connectivity is linked to education level and household income. That raises the chicken-egg question of which came first*—the connections or the intelligence or the education or the income or some combination of these?

On the output end, what would a change in score with any relevance look like? A study that I'll detail below found an effect of an intervention that led to 15-point increases in IQ in the tested population. According to one analysis, that translated into an "upper medium" or "large" effect size.[5] That's the biggest I found. What led to it may surprise you.

As I've already discussed, the relevance of any change will relate in part to where your IQ was when you started. Changes from lower scores carry greater impact than the same numerical change with a higher baseline score. The other question is, what would you get out of it anyway? Nothing here is going to make you Bradley Cooper in *Limitless*, whose character drops a pill and becomes endowed with endless creativity, ambition, and time-management skills. Unlike that character, who eventually manages to retain the effects of the pill without taking it, nothing here seems likely to produce long-lasting, persistent effects. With two exceptions detailed below.

The basics are pretty basic. You can optimize your cognition most straightforwardly and with the least risk by taking what I think of as the Ben Franklin approach: good sleep, good diet, most things in moderation.† Physical activity gets called out over and over as one activity that seems to boost cognition, and caffeine is another.[6]

Odds are, if you're an adult, you're already on the caffeine wagon. But in all seriousness, before you try supplements, brain training, brain zaps, or

---

* The chicken came first, obviously.

† Although nothing will ever prevail on me to graciously accept "early to bed, early to rise."

even learning to play chess (none of which are going to help you much any-way), do yourself a favor and consider taking a daily walk, finding ways to de-stress, and making sure you get enough sleep.

I know. None of that is easy. And you'll notice that no single intervention is considered The Thing that will be this one weird trick.[7] In addition, you have to consider factors like time commitment, how long it takes to experi-ence an effect, side effects (which are neither negligible nor risk free in a lot of cases), what's socially acceptable (you can't be tripping balls at the office party, or at least not the kind I attend), and which aspects of your cognition a given strategy might boost.

If your concern is memory or attention, I deal with those in later chap-ters. I also deal with diet, meditation/mindfulness, and psychedelics in the chapters addressing elements that studies seem to focus most on (e.g., cre-ativity, memory and attention, and depression).

SHARE AND SHARE ALIKE

Stress reduction is important because chronic stress can lead to all kinds of effects in our brains that we don't want, including atrophy, or loss of tis-sue, in areas crucial for learning and cognitive flexibility.[8] One thing that can stress our brains is cognitive overload, in which capacity—especially for things we need to keep in mind and attend to right now—is filled to the maximum.

The theory of a pair of systems working in tandem to keep us from en-gaging in impulsive behaviors posits that one of these systems, the fast one, makes a rapid decision for us, while the other, slow one absorbs that deci-sion, deliberates on it, and can hit override if it seems too rash or danger-ous. Unfortunately for us in these overloaded times, the slow system takes energy—real energy as a resource—to function. If we've maxed the load so that energy is not available, we dispense with the slow, deliberative process and just make decisions on the spot, following the guide of our intuitive, fast system.[9] Researchers test these processes by having study participants hold increasingly long series of numbers in their heads and recall them while also making decisions, such as whether to have a snack now and get a lit-tle money, or have a snack later and get twice as much money. As you can imagine, the greater the cognitive load (length of the number sequence), the more the decisions are affected in terms of risk, impatience, and other factors. If I'm cognitively overloaded, I'm going for the snack now because I don't have time to think about "later."

Our ability to make good choices definitely suffers under a maxed-out cognitive load. If intelligence is the power to effectively problem-solve, as it is often defined, then we are undermining it considerably if we overload the system. That system, studies suggest, includes several regions in the prefrontal and parietal areas, where key hubs in networks relevant to cognition reside.[10]

In fact, when we commit these hubs to maintaining information in our working, or immediately available, memory, areas associated with the default mode system, where we might ruminate more, are less engaged. There is a tradeoff to keeping items in working memory, and the more we load that part of our brains, the more resources are reallocated to support that process. If that sounds like stress, it is.

So how can we reduce our cognitive load and think smarter?

One way is to outsource the load. Neuroergonomics is the study of how we rely on systems beyond our brains to remove some of this load and free up working memory space. Basic examples are keeping a written to-do list or setting alarms on your phone for appointments or important reminders. But machines we use and systems we establish are not our only tools for neuroergonomic adjustments beyond our brains. We also have other people. Not in the "user" sense, and not in a codependent sense, but in a "mutually supportive skills fitting like puzzle pieces" sense.

In my house, that means I serve as the designated executive function, coordinating the schedules and syncing the lives of five people and one elderly dog. That's fine for me because it's one of my skills and doesn't take up a lot of my cognitive reserve (usually). It leaves my partner, who is *not* a fan of cognitive load-bearing to maximum levels, able to do the work that's required of him, which includes intensively focused programming, along with a lot of household tasks and, of course, parenting. I'm an accommodation in that sense, just as he is for me when he deals with the impossibly complex and redundant components of the five remote controls we own to watch one television.

So take stock of your environment and look for ways to adjust it to a better fit for you. The more you can offload to the environment, the more resources your brain can allocate to thinking clearly.

DRUGS: IT AIN'T WORTH IT?

I'm not antipharmaceutical or antidrug by any stretch. People who need help from medications should do what they need to do. But if the goal is to "get smarter" or score higher on an IQ test, I need to point out that across the board, research pretty much indicates that the risks with drugs purported to achieve these goals outweigh whatever benefits they putatively afford.[11] Meanwhile, the public overestimates the benefits while minimizing side effects. Along with that, uptake of "pharmacological cognitive enhancers" has grown considerably, even in the last few years.

Not that we haven't always been at it. Throughout human history, cultures around the world have sought enhancing elixirs. Instead of necromancers, there were neuroenhancers, including caffeine,[12] ephedrine-based potions, nicotine, coca concoctions,* and khat, a plant native to Ethiopia that produces a stimulant compound. These drugs have in common their attention-enhancing abilities because they increase levels of neurotransmitters, such as dopamine, that sharpen alertness to new information. The aftermath and side effects of some of these stimulants, now well known, include everything from the jitters to an irregular heartbeat to sudden death; they even led to recognition of a cocaine "epidemic" at the turn of the twentieth century. (Freud was a huge cocaine enthusiast, recommending it for a variety of brain-related ailments.) Yet these substances did give a cognitive kick and heighten mood, at least transiently.[13] Cocaine, like amphetamine, was embraced for its performance-heightening properties—before its side effects, lasting consequences, and addictive properties became apparent. And there's some question as to whether amphetamines even provide the claimed effects. One study found that participants who thought they'd gotten amphetamines not only self-rated as having better performance in cognitive tests but actually did perform better. The catch is, they'd received placebo.[14]

---

* A performance enhancer from at least the sixth century in South America, and in more recent times, when it was mixed into the original formula for Coca-Cola as a "brain tonic."

As one group of authors put it, these studies "evoke some feelings of déjà vu: A substance aimed at improving human mental and physical performance was used widely before the negative long-term effects and risks became apparent."[15] Even cocaine once was considered helpful for alertness before the negative piggyback effects of this boost became clear. Which is to say, caveat emptor when it comes to pharmaceutical- and supplement-based promises about neuroenhancement.

Today, the popular drugs are prescription modafinil and methylphenidate, the use of which is on the upswing, particularly among students who seek to enhance test performance and gain a competitive edge.* Rates of use among young adults may be as high as one in eight in some Western countries. Evidence that they offer cognitive enhancement is mixed, and the long-term effects in people who don't have the actual indications for them are unknown. See "cocaine" and "amphetamines" above.

Really, the same caveats apply to any intervention targeting your brain in the absence of a clinical indication. In the end, like the fellow in the movie *Total Recall*, you might conclude, "Don't fuck with your brain, pal. It ain't worth it."† And if you're doing it in service to some eugenicist's vision of a number by which to categorize people (meaning IQ), it really ain't worth it.

SUPPLEMENT ROULETTE

Ah, supplements. Growing up, I took them by the handful because it was the 1970s and that's what in-the-know families did. I look back and wonder now what on Earth I was taking, not because the bottles weren't labeled but because of studies showing that they often don't contain what is claimed and often do contain a lot of things you don't want and didn't pay for.

In a study published in September 2020, the surprise contents of the supplements the researchers tested weren't fillers like rice powder or whatnot. Nope. They were active pharmaceuticals, some of them at concentrations fourfold a typical dose—and unwitting consumers are just popping them

---

* These drugs are prescribed for the specific indication of attention deficit hyperactivity disorder, which is a well-studied use, but they are not intended for scattershot use to "boost performance."

† Although he was giving this advice for a somewhat different reason.

into their mouths. Of the ten supplements the authors evaluated, eight of them were alleged mental enhancers. Yet in all eight, an unapproved and unlisted drug was the key ingredient.[16] The safety profile of these drugs is not clear, but as reported by *STAT*, alarmingly, one of them carries a warning that pregnant people shouldn't use it.[17] Unfortunately, the US Food and Drug Administration has little in the way of tools to regulate the supplement industry, thanks to a congressional act passed in 1994 that prevents the agency from evaluating supplements in the manner of pharmaceuticals.* Thus, the consumer is left to decide if the claims are true and then to risk exposure to undisclosed chemicals, all in the name of "cognitive enhancement." They are playing a roulette of sorts with fifty-fifty odds. As the authors of the study wrote, "Several detected drugs were not declared on the label, and several declared drugs were not detected in the products." Furthermore, for supplements that did contain the declared products, 75 percent of them did not contain quantities that matched the label.

NEUROHACKING OR NEURORISKING?

Brain-computer interfaces, or BCIs, which I introduced in the second chapter, could be considered a form of neuroergonomics. After all, they are systems that serve as extensions of our minds, and in some cases they explicitly act in that way.[18] This section doesn't deal with mind-melding and such. Instead, we're going to look at what neurofeedback and brain stimulation from the outside do to cognition.

As with almost all the other interventions described in this chapter— besides those that simply promote overall salutary practices (physical activity, good sleep, healthy diet)—caution is warranted with these approaches in the absence of current indications, such as intractable depression in the case of transcranial magnetic stimulation (TMS). Certainly, research suggests that these interventions affect the connectivity infrastructure of our brains.[19] The ratio of benefits gained versus costs incurred remains unclear. That is a

---

* Apparently, plans are in the works to do something about that, including the introduction of an online tool that is intended to provide consumers with information about ingredients that "do not appear to be lawfully included in products marketed as dietary supplements" (www.fda.gov/food/dietary-supplement-products-ingredients /dietary-supplement-ingredient-advisory-list).

bit worrisome given that a lot of transcranial futzing around with the brain has become a do-it-yourself endeavor, aka "neurohacking." To remind the reader, the Planet Brain is not a computer to be "hacked" any more than planet Earth is.

The most engaged DIY community centers around transcranial direct current stimulation, or tDCS, which you read about in Chapter 2. A trip around the online world of tDCS devotees was enough for me to know that I wouldn't touch the procedure with a ten-foot pair of insulated gloves. Comments on people's real-world experiences range from someone who ran "enough current to burn and create red marks on forehead" to "I started tDCS 2 months ago. Since then I have had real trouble with spelling words, I find myself—for the first time in my life—having to google how to spell even the most simple and everyday words." An anti-testimonial, if you will.

For more scientific observations, to quote the authors of one recent review, "Commercial do-it-yourself electrical brain stimulators might impair rather than enhance cognition."[20] As the authors of another review note, "Its efficacy and even effects on brain and behavior are currently disputed."[21]

Among those disputed findings are early hints that tDCS could boost some problem-solving skills, including those needed for a verbal task that involved finding a common word associated with three cue words.* This test, the Remote Associates Test, was invented in 1962 to measure creativity. Another study, from 2012, this one using a different test, found that four out of ten participants who were zapped in the temporal lobes with tDCS were able to solve a difficult puzzle† that no one could solve who hadn't undergone the stimulation.[22]

As a chaser to that 2012 study, in 2017, a group that tried a similar setup—inhibiting the left anterior temporal lobe and stimulating the right, with a

---

* Here's an example I just made up. The cue words are "pasta," "Waldorf," and "bowl"; the common word is "salad." Try it for yourself here (but without DIY brain zapping, please): www.remote-associates-test.com/.

† The task asks the participants to connect nine dots in a 3 x 3 array with four straight lines, drawn without lifting pen from paper or retracing a line (Figure 4.1). The authors claim that a century of research has established that "the expected solution rate for this problem under laboratory conditions is 0%. . . . It is known to be so difficult that most people fail to solve the problem even if they are given hints, a long time to solve the problem, or 100 attempts to solve the problem."

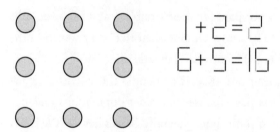

Figure 4.1. A pair of brain teasers used in tDCS studies, as described in the text. Solutions appear at the end of the chapter.

sham control—did not show improvements in resolving the "matchstick" arithmetic test. In this test, participants make a math statement "written" in matchsticks correct by moving one of the sticks, and only one, to another position, without adding or throwing away any sticks (Figure 4.1). The researchers also administered a culture-specific verbal-insight problem (using Japanese kanji). As a control task, they had participants solve easy two-digit arithmetic problems. They found no differences among the groups.[23]

Yet another study used only right temporal lobe stimulation and a different kind of task (the Remote Associates Test described above). These authors found that participants in the stimulated group improved over their pre-stimulation baseline and performed better than the sham-control group or a group stimulated in the left frontal cortex.[24]

Meanwhile, in 2020, still another group targeted a completely different region, the right posterior parietal cortex, for its associations with spatial reasoning.[25] They concluded that stimulation here "significantly enhanced" the ability to solve indeterminate deductive problems. By that, they mean taking a series of sentences explaining the relative location of items in pairs, and then sorting those items in the correct order.*

As is common, each of these studies used a different protocol, design, and test, making comparisons among them or drawing any firm conclusions a tad difficult. Furthermore, none of the studies involved huge numbers of participants—twenty-two split between two groups in the 2012 study with the nine dots, sixty-six split among three groups in the Japan study ("all in

---

* Example: The duck is to the left of the toucan. The eaglet is to the right of the toucan. The loon is to the left of the warbler. The loon is to the right of the toucan. Is the following correct? Duck—toucan—eaglet—loon—warbler. See the caption under Figure 4.2 for the answer.

return for money," as in they got paid for their troubles), and fifty-one split among three groups in the 2020 study. Despite their size, they were all otherwise carefully designed (I hope), controlled lab studies.

One approach that has been proposed is to couple tDCS with "cognitive training."[26] So far, efforts have focused on older populations and not much on the transfer of gained skills more generally. That's key. While it has come to be expected that people might experience improvement in the skills they're using for these games and tests, known as near transfer, the brass ring here is far transfer, or generalizing to unrelated abilities because you've become globally smarter.

MAGNETIC MINDS

TMS is also on the radar for cognitive enhancement, but this application doesn't lend itself so much to DIY. Its lag behind tDCS in the DIY world is suggested by its still having been viewed in 2014 only as a "promising alternative" to pharmaceutical-based approaches to enhancement.[27] One reason may be cost—the magnets and imaging techniques associated with the practice are not something you can just order off eBay.

Like tDCS, TMS operates by exciting or inhibiting activity in specific areas of the cortex that are thought to be important to cognition—and thus possibly enhancing it. The stimulation frequency determines whether the pulses are exciting or inhibiting, with low frequencies of about 1 Hz decreasing excitability and higher frequencies of 5 to 20 Hz increasing excitability.

Despite the obvious appeal to being able to make adjustments merely by tweaking pulse frequencies, results of early efforts to use TMS for cognition were not especially Earth shattering. The authors of a 2014 review discussed finding sixty-one "instances of performance enhancement" linked to TMS, but the enhancement was in subcategories, such as attention, memory, and object identification.[28] These interventions have mostly been performed in people with some kind of known brain compromise, such as stroke, memory impairment, or traumatic brain injury.

What also remains to be seen is how real any of the effects are. TMS is not easily shammed because the patients sense vibrations through the coil and can respond to that alone—or can understand when they are absent.

For this reason, actual TMS compared to sham TMS that produces vibrations could be attributed to, as those review authors put it, "general arousal."*

A 2020 study of twenty healthy young participants offers a tantalizing (if this is your thing) hint that TMS could boost fluid intelligence. Using an advanced TMS protocol that relied on individualized imaging data, the authors found that their approach, called cortico-cortical paired associative stimulation, or cc-PAS, yielded accuracy improvements in tasks related to fluid intelligence. If results were to be replicated, they'd have a winner on their hands.

The "pairing" in this method involved applying the TMS stimulation as a paired pulse to two cortical regions, hence the "cortico-cortical." The approach was expected to strengthen the synaptic connections between neurons, which in turn was expected to translate into being able to use them more easily—or fluidly, as it were.† More specifically, stimulation in the parieto-frontal direction (back to front) was linked to improved logical reasoning (the reverse *decreased* it), and stimulation in the fronto-parietal direction (front to back) was linked to improved relational reasoning.[29] Yet the authors didn't conclude that everyone needs to rush out and perform TMS on their brain. Their aim seems to have been more research oriented, as in making a discovery that could help researchers be more selective in targeting brain regions in the context of what they are studying.

In any case, the effects in the 2020 study were transient, which has been a common finding with TMS studies, with some effects lasting only a few seconds or minutes—or a couple of hours in other cases. One report cited eighteen solid hours of improvement (enough time to read *War and Peace* at top speed or something).[30] The brain seems to snap back after attempts to stretch it. One explanation is that homeostasis reorganizes everything back to the way it was as a defense. Furthermore, researchers report having found no transfer of these transiently gained (or lost?) cognitive abilities to other kinds of cognitive abilities.

---

* Not of the sexual kind but of the "I have heightened awareness" kind.

† This prediction arises from a theory of how learning might work, in which the more a neuron stimulates another neuron, the faster/better/smarter the operation of that communication network will be; it's called Hebbian plasticity.

Another group examining repetitive TMS (rTMS) in a 2020 study found performance improvements only in the most difficult working-memory tasks, and not much of an overall effect of the intervention.[31] These authors wrote that the idea that stimulating a single cortical node could yield lasting change is "still highly debated." However, they concluded that the results triangulate control of a brain network, activity in the brain (as determined from imaging), and behavior (performance on the tasks), and support the idea that a brain network and its activity could indeed be targeted.

LET'S GET PHYSICAL

I know that everyone hammers on exercise as the true one weird trick for all the things, but in fact, evidence backs up this assertion. Physical exercise is linked to cognitive improvements, whether directly or indirectly.[32] Some authors go so far as to identify it and diet as the only confirmed interventions that have lasting effects. Exercise is, as one group wrote, among the "cognitive enhancers with the most wide use and longest history."[33] They added, "A rapidly growing body of evidence shows that everyday activities . . . such as physical exercise improve cognitive functioning." That's the most definitive statement I found in hundreds of papers on the subject of cognitive enhancement.

Certainly, the directly beneficial effects of exercise on the nervous system are clear. Your frontal region gets an oxygen infusion, your cortex is activated, your neurons metabolize more efficiently.[34] I think it matters—and I hope it makes a difference—to know what happens on the ground in Planet Brain, rather than just hearing, again, "You need to exercise every day."

On the molecular scale, a protein called brain-derived neurotrophic factor has a strong role in brain plasticity—the flexibility to build new and strong connections. Exercise positively affects levels of this protein and of neurotransmitters known as catecholamines (e.g., dopamine, epinephrine). In turn, these signaling molecules drive flexible shifting of attention from one thing to another, variously called "task switching" or, in a more specific subset, "set shifting."[35]

Flexible shifting is associated with cognitive fluidity. As you may have sensed, "fluidity" is a key word in cognition. Set shifting has been associated with problem-solving capacities and—befitting an overall theme of this book, and possibly your experience at cocktail parties—with social abilities too.

A meta-analysis performed in 2021 of twenty-two studies confirmed the benefits of acute exercise on set shifting in nineteen hundred total participants. The effects were more pronounced in older versus younger adults (who are probably just moving around more anyway), and light-intensity exercise seemed to be the most beneficial. However, the effect was not large, and the authors say that the "relevance for everyday life is questionable."[36]

I don't find that statement overwhelmingly compelling, or I wouldn't if I didn't already believe in the overall value of being physically active. But people are willing to gulp supplements, zap their brains with DIY electric contraptions, and undergo magnetic stimulation, with all the attendant risks, for the teeniest transient effects—if any. Exercise, with its many physical and mental salutary effects, earns a place in line ahead of these interventions.

But hark: there's more. We can add "enhanced executive function" to "improved set shifting." A systematic review of thirty-six randomized, controlled trials of 4,577 students (yes, young people) found that physical exercises were associated with significantly better working memory, inhibitory control, and cognitive flexibility.[37] The effects on inhibitory control and cognitive flexibility were less notable than the effect on memory. But layering the numerous benefits of exercise that have been identified in these meta-analyzed bundles of controlled trials adds confidence and support to the notion that of all the individual, accessible interventions being discussed in this chapter, this one has the biggest collective effect.

Still not sold on even minimal daily activity? Two other meta-analyses examined how exercise affects executive function. One evaluated data from thirty-three randomized, controlled trials and found an association between exercise and improved executive function in adults, with greater effect sizes for moderate- versus low-frequency exercise, and for younger (age fifty-five to seventy-five) versus older (over seventy-five) adults.[38] These authors looked at different types of exercise and found good results with yoga and tai chi.

The second meta-analysis, which looked at randomized, controlled trials of adults age sixty or older, found that engaging in exercise versus doing no exercise was associated with better working memory, cognitive flexibility, and inhibitory control.[39] You may notice a distinct pattern here. Different types of exercise, such as aerobic versus "mind-body," offered different magnitudes of influence. But in both meta-analyses, the good news is that every kind of exercise seemed to be associated with some effect—so you can choose what you like. Just choose something.*

MINDFULNESS

You'll have noticed that one aptitude researchers test over and over is "cognitive flexibility." That's because this capacity, which arises from intact executive function, produces our solutions to problems, even new ones, along with our ability to move fluidly between tasks. When it comes to thinking, rigidity and flexibility are at odds with each other, just as they are linguistically, and for cognition, the emphasis is on flexibility.

In a randomized trial of the effects of mindfulness-meditation training on a hub in the executive control network, thirty-five "high stress, unemployed, job-seeking adults" engaged in either three days of intensive mindfulness training† (squeezing an eight-week program into that short time frame) or three days of relaxation training as a control.‡ Researchers then compared the participants' baseline neuroimaging to their postintervention imaging, performed within two weeks of the end of the training. They found greater increases in connectivity among the hubs of the executive control network following the mindfulness training compared to the relaxation training.[40] We never get to just sit around and be magically improved, do we?

CAN GAMES MAKE YOU SMARTER?

At least maybe we can achieve the gains we want by playing some fun games, like Tom Brady, right?

No.

---

* This is just about the only direct, intervention-positive advice I will give in this book.

† "Body scan awareness exercises, sitting and walking meditations, mindful eating, and mindful movement (gentle hatha yoga postures)."

‡ They named their programs Health Enhancement Through Mindfulness (HEM) and Health Enhancement Through Relaxation (HER). Get it?

The bottom line with all the brain-training games out there is, as Daniel Simons put it, "If you practice something, you'll get better at that task." That's it. So if you engage in "brain training" that has you clicking at ghost images of butterflies on a screen, and you do it enough, you'll get better at doing exactly that thing. That's "near transfer," an effect right where you're "training" your brain.

What you won't experience is "far transfer," in which that near effect might generalize to some other cognitive skill, or even go global. Yet the marketing departments at these companies tell us that they have dozens or even "more than a hundred" studies supporting the effectiveness of what they are selling. As discussed earlier in the chapter, salami slicing can be one reason that stacks of studies can look so bulky.[41] Publication bias (see Chapter 2) is a problem too—an attempt to hide the salami, as it were, rather than slicing it.

Another issue is the lack of active controls. A control group needs to be actively doing something that captures all aspects of the intervention itself except its potential cognitive effects. Without this, any perceived influence could be related to all kinds of confounding variables, like general arousal because clicking on ghost butterflies is fun.* "There is no evidence that these will translate to the real world," says Simons, which seems valid unless your real world consists of chasing digital butterflies.

Giovanni Sala, who probably knows his way around these studies as well as anyone, is even more blunt. Evidence for "far transfer induced by training on cognitive tests seems to be simply zero," he told me. He would know because he has dug into every single data point out there. He's even unearthed data tucked away in supplemental materials, the often-overlooked adjuncts to peer-reviewed publications. The supplement is where some authors might, you know, slide in results so they can at least claim that all the data were there, transparent for everyone to see . . . if anyone ever looked.

Sala and his coauthors, including Dan Simons on one meta-analysis, did look. At everything. And found nothing. The careful analyses performed by

---

* It is not.

Sala and his colleagues showed again and again that when reported effects were corrected for placebo effects, there was no there there.[42]

Another group, Kühn and colleagues, who specifically assessed the potential of video games to exert some benefit on cognitive performance, noted that "the devil is in the details."[43] An examination of cognitive targets, genres of games, intensity of training, and other factors simply showed that "results are mixed." They wrote, "It seems almost impossible to outline the effect of video gaming on cognition in a simple statement without mentioning the numerous limitations." You can almost feel their frustration at the lack of consistency in anything about this area of study.

The debate has been heated, and it continues: a meta-analysis of video game data that Sala published with a co-author made note of an angry back-and-forth between academics and owners of a website maintained by the founder of one of the brain-training companies.[44] Their competing conclusions could not be more stark: the academics, at Stanford and the Max Planck Institute, wrote that "there is no compelling scientific evidence" that such games reduce or reverse cognitive decline, and representatives of the industry-hosted site offered the riposte that "there is, in fact, a large and growing body of such evidence." With such claims, it's good to determine whether or not that body consists of slices of salami.

"You should play these because you like them, not because you expect to get better at something else," says Sala. And he extends that to other putative "brain-improvement" activities, like chess and learning music. "There are many reasons you should play an instrument or play chess or play video games," he says, "but getting more intelligent, in my humble opinion, is not one of them."

"Music!" you exclaim. Nope. Not even learning music. Sala and a colleague looked at this too. While acknowledging that postulating a link is "natural," given the common association of highly talented musicians with intelligence and creativity, they found only a small effect of music education in childhood. If they considered only studies that compared musical training with an active control, they found effects "close to zero."

Their conclusion is that although a musician's global cognitive ability might correlate with their musical aptitude, it does not correlate with years of training. "Intelligent people are more likely to engage and succeed in the field of music," they wrote.[45] Or people of higher socioeconomic status, which is closely entwined with scores on measures of intelligence, can afford to pay for music lessons for their interested children, I write.

And no, not chess, either. Checkmate, cognitive enhancement. At best, chess has a small to medium effect on cognitive skills, but when you confine analyses to studies that compared chess training to an active control of some other activity, there appear to be no treatment effects at all.[46] Sala, who himself is rated as a chess master and has taught chess, says, "Just do it for the sake of it."

The stakes are high. As another researcher told me, being able to make people smarter "could be a Nobel Prize–worthy achievement" because it means changing people's lives for the better. "It's a real bummer, a very real bummer, for many people that it's not possible at the moment."

THE SOCIAL INTERVENTION

Poverty interferes with a lot of things, and among them is cognitive function.[47] The stress of poverty and the overwhelming attention it takes pull resources away from the careful, monitoring parts of our minds. And of course, it's difficult to be mindful or meditate in the absence of stable sources of income, food, and housing.

One study of farmers in India and shoppers at a New Jersey mall—yes, in the same study—found that people whose socioeconomic status is low experience difficulty with decision-making under the trigger of having to think about a significant financial outlay. They did not have this issue if the outlay was minimal. Well-off people did not experience this cognitive impedance under either condition. And among the farmers, for example, the interference was present before the harvest, when poverty was at its worst, but also after harvest, when they were comparatively well off.[48] A lifetime of that kind of burden might leave no breathing room for leveraging the plastic potential of the brain to reach its full abilities, and poverty is the factor limiting that breathing room.

Of all the strategies we could adopt to improve people's cognition and win a Nobel Prize, lifting all economic boats seems to be it. Unfortunately, a lot of us are not so willing to help. Or they can't.

THE GENE-GINI INTERPLAY

A 2019 study of social inequality found that it unsurprisingly has a "profound effect" on what we measure as the heritability, or contribution of genetic variants, of educational attainment. In fact—and oh so intriguingly—the authors wrote that the more equal the members of a population are, the more heritability accounts for educational level achieved. A level economic playing field seems to erase environmental effects, suggesting a large role for economic equality in environmental influences. This pattern, the authors say, indicates that inequality suppresses an ability to manifest behaviorally the gene variants related to education level. As the poverty study indicated, the social factor of inequality shuts down potential.

The study authors—Fatos Selita of Tomsk State University, in Russia, and Yulia Kovas of Tomsk and of the University of London, UK—call this interaction of gene variants with an equality measure known as the Gini coefficient* the "Gene-Gini interplay."† Inequality, they say, affects us from the fetal stage to birth and throughout our lives. It has been linked to a host of problems in childhood, including lower education and an education achievement gap.[49]

That gap translates into a growing chasm between children from low-income families and wealthier peers in proxy measures of ability, such as standardized achievement tests. From 1970 to 2013 in the United States, for example, the gap between children from families in the ninetieth percentile of income and those from families in the tenth percentile expanded by 40 percent. Interestingly, as the inequality factor decreased somewhat between Black and White children from high levels in the 1950s and 1960s, the achievement gap between those two populations began to shrink too.

What's strange and worth noting is the regional pattern. The heritability of intelligence in the United States interacts with socioeconomic status, and intelligence scores track lower with lower socioeconomic status and higher with higher status. The thing is, this interaction isn't found in Western Europe or Australia.

---

* After the statistician who developed it in 1912, Corrado Gini.

† My fervent hope is that this is a reference to the David Bowie song "The Jean Genie."

What is the difference between those countries and the United States? It's not socially defined race, because the interaction holds even when controlling for that factor. What does stand out is a higher wealth and income Gini score in the United States, reflecting a wider wealth and income inequality gap, and more people living below the poverty line. As Selita and Kovas put it, "The Western Europe countries and Australia have more uniform access to high-quality education, health care, overall social welfare and greater social mobility than the US."

The high level of inequality in the United States sets the stage for those experiencing its negative influence in every aspect of their lives, from their health—as the COVID-19 inequalities in infection rates, severe disease, and deaths show—to their access to education, with related effects on limiting their cognitive potential. Selita and Kovas wrote, "A disadvantaged environment may reduce genetic effects because it lacks 'food' (educational opportunity, resources, encouragement, choice of activities) for children's genetic 'appetite' for learning." This starvation of potential can put the brakes on cognitive and language development, dictating a lifelong trajectory and stifling the expression of educationally relevant genetic propensities.[50]

But can education really do anything?

As Suzana Herculano-Houzel has written, "The difference is that having enough neurons endows a brain with the *capacities* for complex cognition, but turning those capacities into actual abilities takes a lifetime, if not generations, of learning." Through learning, "the abilities developed are passed on and accumulated. This is the case to this very day in our modern, complex, busy lives."[51]

As she notes, regardless of the *capacity* with which we are endowed as a species or individually, *learning* is what fulfills that potential. She writes, "The difference between capacities and abilities is made painfully clear in the extreme cases of children who are raised in near solitude, or with hardly any opportunities for learning. . . . Their brains remain arrested at the stage of promising raw material that could achieve much—but doesn't."

In a sweeping analysis of various cognitive interventions, Sala and his coauthor Fernand Gobet found only a marginal enhancement of global cognitive ability with education.[52] But as we've just seen, inequality can starve even the heritable component of cognitive ability. Annette Brühl, at the University of Cambridge, and colleagues wrote that it's reasonable to argue that an absence of cultural or social enrichment is "one of the main reasons for the well-known cognitive and emotional developmental deficits due to low socioeconomic status."[53] They even noted that we tend to view these things as "circumstances" and not as "interventions," but their effects are "widespread." "From an empirical descriptive perspective," they wrote, "one could say that low and negative socioeconomic circumstances are toxic for cognitive development or performance and that the beneficial effects of high socioeconomic status are comparable to neuroenhancement."

People have attempted to test this hypothesis directly, including in a couple of well-known preschool projects begun in the 1960s and 1970s.* These projects had their flaws,[54] like all studies involving countless unmeasurable factors, especially in public health. But they certainly suggested that targeted programs to reduce the toxicity of poverty, inequality, or low socioeconomic status have effects on health, life trajectory, and, yes, intelligence scores and educational achievement.[55] It's education as antidote.

Here's a series of premises: We view intelligence as the general ability to solve problems.[56] (True, no one can agree on what intelligence is, but just work with me here.) We've found that most alleged brain-boosting interventions don't have far-transfer effects but can enhance proximal skill. We know that childhood is a malleable and vulnerable period, and that environmental factors can squelch the contribution of genetic effects. We know that of all the interventions that do anything, educational attainment[57] and physical exercise top the list. Doesn't it seem like to win that Nobel Prize, we need educational programs in preadulthood that focus on the general ability to solve problems, achieving near transfer for that skill and thus getting at the core of intelligence?

For every year of additional education, IQ measures increase by 1 to 5 points, according to a huge meta-analysis of data from six hundred thousand people.[58] The authors of this study concluded that "education appears to be the most consistent, robust, and durable method yet to be identified

---

* The Abecedarian Project, begun in 1972 at the University of North Carolina, Chapel Hill, and the Perry Preschool Project, begun in 1962 in Ypsilanti, Michigan.

for raising intelligence," with effects lasting a lifetime, rather than seconds or minutes or hours. They also point to some evidence that the reason education does this is because it has "an equalizing effect." Gene, meet Gini.

THE BELGRADE STUDY

In 1956, Lazar Stankov, now a professor emeritus at the University of Sydney, was in high school in a small town in the former Yugoslavia, where, he told me, "communist rule was a little bit less pronounced because of Tito." People were getting more leeway to teach at universities, and the universities were gaining a little more leeway in what they could teach. One of the new areas was educational psychology.

Radivoy Kvashchev took a course in educational psychology at the University of Belgrade, his home region, and was so taken with the field of study that he enrolled in a PhD program. Just at that time, an enthusiasm for educational psychology had taken hold in the grade schools too. Kvashchev was assigned to a local high school that Stankov attended.

Stankov met Kvashchev, "talked to him a lot (I was an academic baby at the time)," and listened, rapt, as the older man discussed the idea of increasing IQ and expanding students' abilities. So taken was Stankov with these notions that after he graduated he enrolled in psychology at the University of Belgrade. There, he quickly found his attention drawn by the allure of the statistical muse, which once again carried him into Kvashchev's sphere.

While at Stankov's high school, Kvashchev had implemented a multiyear program intended to teach the students how to think critically and solve problems. By the end of it, "he had a huge amount of data," Stankov related, and he needed someone who understood statistics to deal with those data. "We were using those cranking little calculators," recalled Stankov. Thanks to help from Stankov, Kvashchev completed his PhD thesis and went on to become a professor of educational psychology, publishing eleven books in Serbo-Croatian before his untimely death at age fifty-three in 1983.

The results that Kvashchev obtained sufficiently impressed the US developers of the intelligence, creativity, and critical-thinking tests he used that they trekked to Yugoslavia in 1968 for a meeting. What was so impressive? "His work was showing an effect of the intensive training," said Stankov. "In the course of the training, you are able to change IQ." You will note that the training occurred in a school, lasted for three years, and, as you'll see below, was both engaging and on target for problem-solving.

Because many people were not reading papers or books published in Serbo-Croatian, Stankov felt the need to "bring this out into the wider world and point out that this was important." Furthermore, he did not hesitate to add, he felt that in publishing the findings, "some of these genetics arguments would be diminished." He wanted to produce this evidence against hereditarianism.

Stankov the statistician and psychologist bumps up a bit against the Simonses and other colleagues of this field, who report having yet to see any far transfer of training on cognitive abilities. Stankov himself published work in the 1980s suggesting that practicing two cognitive tasks at once was associated with improvement not only in the components of the tasks but also in that elusive g, or "the common variance" that tests of fluid intelligence are thought to capture.[59] And in the Belgrade study, the target was problem-solving skill, so even if the effects were limited to near transfer, they would still go straight to the core of what we view as intelligence.

Stankov and his coauthor, Jihyun Lee, from the University of New South Wales, found the conclusions of Simons and colleagues "rather pessimistic." But, they argued, the three-year training study that Kvashchev completed in Yugoslavia addressed all the shortcomings that Simons et al. highlighted— small numbers of study participants, short time frames, limited measures of intelligence, and lack of randomization. In fact, Kvashchev's study involved hundreds of students, lasted three years, had a control group, and used more than two dozen measures of intelligence at four testing time points.

In an overview of the literature on education and IQ, Stankov and Lee summarized the findings associating education with directly increasing intelligence scores. These patterns were replicated with strikingly similar outcomes in a 2020 study in Denmark, which showed that education drives intelligence scores higher and has a greater effect in children with lower baseline scores at age twelve.[60] Another meta-analysis of 111 studies demonstrated links between problem-solving creativity and education-related factors, such as teacher encouragement and teaching approaches, which are social effects.[61] Specifically, the approach with the strongest effect on problem-solving contained the component of *defining the problem*, including coming up with as many ways as possible of solving it.

Stankov and Lee were motivated by these findings and by how Kvashchev's study design addressed the limitations of many studies, along with the fact that Kvashchev had used a "defining the problem" approach. They undertook a reanalysis of the results of that three-year program in Belgrade. The process of problem-solving it relied on requires three elements that are also involved in the work of the DMN, or the "sense-making network," as some researchers recently proposed.[62] These elements are (1) detecting what is happening, (2) placing it in context of what came before it, and (3) recognizing what our place is in the situation and how it's relevant.

The Belgrade program consisted of first-year high school students in two schools: 147 in five classes in one school, serving as the control, and another 149 in the second school, also across five classes. In both cases, the number of students represented about half the student population in the grade.

The control students attended school as usual. The students in the intervention group had classes at least once a week in creative problem-solving, taught by teachers whom Kvashchev had trained to create appropriate exercises for their subject area—language, math, or science.[63] The exercises had to meet specific criteria, including drawing in factors that were not closely related, requiring students to "radically" reformulate and reorganize the information before they could solve it, and eliciting both divergent (connecting apparently disconnected ideas) and convergent (funneling toward a consolidated solution) thinking. In addition to promoting problem-solving, the program promoted creativity, a rare feature of interventions intended to boost IQ.

Here's an example of one of the problems:

I was captured by a band of outlaws—said a famous explorer—and their leader had my hands and legs tied up so that I could not move. They did not gag me up, though, and I was able to use my mouth freely. The leader of the gang hung a piece of bread exactly five centimeters away from my mouth. He then laughed and said: "If you manage to eat this piece of bread, I'll set you free." He knew that I could get no help. Also, in order to ensure that I could not roll over or move closer to the bread, they tied me to a tree. Nevertheless, I managed to free myself. How?*

Students had to list as many solutions as they could conjure. What would you come up with?†

All participating students took twenty-eight intelligence tests on four occasions: before and at the end of the three-year intervention, and then two retests, one at the beginning and one at the end of their final year of high school, during which they received no problem-solving training. The tests were all subtests from verbal and nonverbal IQ tests, and the problem-solving training was not designed as practice or preparation for them. Six of them were tests developed by Raymond Cattell.

Stankov published an initial analysis of Kvashchev's study in 1986.[64] In that analysis, using the average performance across all twenty-eight tests as indicative of global intelligence, Stankov crunched out the associations. The control group students at age fifteen outscored their treatment group peers before the intervention. At the end, the treatment group scored higher, and that difference intensified at both retests. But they didn't outscore on all twenty-eight tests, only on twenty-six of them.

In IQ terms, Stankov's first analysis showed that the treatment group members had IQ values an average of 5.66 points higher than their control group peers. At the second retest, the experimental group students had an average IQ that was 7 to 8 points higher. The effect was "medium," more than just about anything else with myriad other interventions, which clock in at best with "small" effects. This does look like far transfer, with a single intervention targeting a skill having a much broader effect—unless you view it as near transfer of problem-solving, a core skill for intelligence. But when Stankov wrote about the results in 1986, he still felt that a question

---

* This happens to be very much like a common test of creativity, as we'll see in Chapter 9.

† Some proposed solutions: blowing at the bread to create a pendulum, or assuming some particularly favorable conditions, such as the wind blowing and moving either the rope or the tree to which the explorer is tied (or both).

mark lingered over them, so he titled his paper with a question, "Can We Boost Intelligence?"

Stankov and Lee returned to the data in 2020 for a reanalysis, erasing the question mark in the process. The original result, they argued, was extremely conservative, having been calculated using higher values of variance from the first testing. When they recalculated the findings using the tighter values for variance from the second retest, they found a difference of 14.28 IQ points between the two groups, for a much stronger effect, one that is "upper medium" or even "large."

Even at the 7- to 8-point difference, the real-world implications could be meaningful, given the associations of IQ with lifetime outcomes. At double that, they undoubtedly would be. In some analyses, the difference was as high as 15 points, some of it in fluid and some of it in crystallized intelligence measures. Stankov and Lee finally settled on a conservative value from the work, saying that the experimental group gained "at least 10 IQ points more than the control group at the end of 4 years." The program seemed to have allowed the students to match their abilities to their capacities.

Kvashchev did undertake a few more training studies with "similar outcomes," but the data were not available for analyses after his death. Stankov told me that Kvashchev's favorite part of the entire experience was "seeing how cute some of the [students'] solutions" to the problems were. If they were struggling, said Stankov, Kvashchev might give them a little hint, starting with, "Some people did say that you might be able to do it this way." The students "loved it, as you can imagine."

It's possible that one of the things they loved was the positive social attention they were getting, including what sounds like a series of fun, creative interactions with each other over the years. The social effects of education are entangled with the pedagogical influence, and studies show that social interactions boost cognitive performance across ages.[65] As a species, we tend to educate in groups. In fact, when millions of US schoolchildren had to participate in virtual schooling during 2020 and 2021, the related hand-wringing was as much about the loss of in-person interaction as it was about the logistical and other problems with distance learning. Perhaps intuitively, we knew that the losses went beyond the realm of scholastic attainment.

Part of education's antidotal effects against the toxicities of inequality may well trace to the social benefits of in-person schooling.

Justin Sanchez is a former director of the US Department of Defense's Defense Advanced Research Projects Agency, or DARPA. Because this agency pushes the bleeding edge of brain-technology research, you'll see more about it in later chapters. But here, I wanted to highlight something that Sanchez, who led the vanguard of the vanguard of brain-enhancement programs for years, has to say about education. He views using the fancy technologies in development to enhance memory as a version of education, and he told the *Atlantic* journalist Michael Joseph Gross in 2018, "School in its most fundamental form is a technology that we have developed as a society to help our brains do more."

CHECKLIST

To use a metaphor from the olden days of brain comparisons, in this chapter we have filtered out many of the interventions that are promoted—without much evidence—as boosting individual cognition. But we have also identified a few evidence-based options for enhancing our brain power, including physical activity and reducing our cognitive load. Here's a checklist of scientifically supported ways you can help your brain help you:

- ✓ Did I engage in physical activity today?
- ✓ What neuroergonomic tools can I identify that might bear some part of my cognitive load?
- ✓ What tool (digital or other) can I use to keep lists?
- ✓ What tool can I use to send me alerts or reminders?
- ✓ What processes have I made too complicated that I could simplify?
- ✓ What procrastinations do I engage in that create a greater psychological burden than just doing the task?
- ✓ In what ways can my social networks make mutual and healthy accommodations to share cognitive load so that we all have a lighter burden?

✓ Remember that "no" is a complete sentence.
✓ Probably don't order a DIY dTCS kit.

We also have seen that on a larger social stage, reducing inequalities and targeting problem-solving educational programs can have big (or maybe medium) effects. Where there's political will, there may be a way to apply these tailoring tools to shaping young minds so they can be expressed to their fullest measure.

In the next chapter, we'll take a look at how social cognition itself might be improved, another inroad to recognizing inequalities and being smart about meaningful connections.

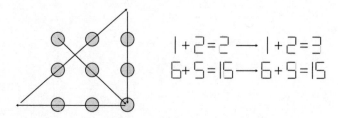

Figure 4.2. Solutions to the puzzles shown in Figure 4.1.
Bird-order puzzle solution: Yes, that is the correct order.

CHAPTER 5

SOCIAL COGNITION: TAILORING THE BRAINS

IN THE LAST chapter, we landed on long-term problem-solving practice in a learning environment as an effective tactic for meeting our cognitive potential. Part of that effect may come from the influence of simply being around people—who, let's admit, sometimes also present problems that we must work out.

THE ME BOOK

In fall 2010, Ole Ivar Lovaas, who had become famous for developing a "therapy program" for autistic children, died. His passing led to a series of hagiographic tributes, one saying that the "field of autism treatment and behavioral analysis suffered a significant loss."[1] His contribution, according to this particular postmortem, was that he flouted "unsubstantiated conventional wisdom" and changed the course of autism therapies.* He showed,

---

* Which he did, just not in a positive way.

it was said, that "some children with autism can make significant progress with early intensive behavioral intervention."

That "early intensive behavioral intervention," most people would agree (I hope), was abusive and sadistic and has caused immeasurable harm to autistic children subjected to it.

A technique used to train rats couples a non-noxious stimulus with a noxious one, specifically electroshock floors. With a goal to use this "operant conditioning," Lovaas modeled his "behavioral intervention" for autistic children on this method invented for rats. A reward would be offered for a desired behavior, and a punishment—including physical punishment and, yes, electroshocks—would be offered for "bad" behavior. "Improvements" in the targeted behaviors often had little social utility. They simply were focused on "training the autism" out of autistic children. And he made little to no effort to use social cognition to understand what might underlie their "unwanted" behaviors, such as "tantrumming."

Lovaas, among his personal achievements, in 1981 published *The Me Book*, a guide purported to lead therapists or caregivers through achieving behavioral adjustments in children who "suffer from mental retardation, brain damage, autism, severe aphasia, emotional disorders or childhood schizophrenia." From *The Me Book*, a small industry arose around a concept called applied behavioral analysis, or ABA. It comes complete with required certifications of people who administer this "therapy," its own scientific journals, and a lobbying arm powerful enough to ensure that ABA is the *only* therapy for autism covered by insurance in the US. It is impossibly costly otherwise, in more ways than one.

The book opens with a preface describing its approach as a "set of programs" meant to help parents and teachers "in dealing with" children who have autism. It earned its title, the preface explains, because the programs, if followed, will ensure that "the child does become more of a person, an individual, more of a 'me.'"

The system Lovaas developed relies on giving "rewards" to children who comply with pointless tasks or demands, such as avoiding flapping their hands. For autistic people, hand flapping can be a release valve for excitement or anxiety, somewhat like self-grooming by brushing or flipping hair around a finger, and it harms literally no one. For their compliance, children were to be rewarded incrementally.* Initially, these increments translated into a "one-second lick" on a lollipop, or a swallow of juice. Why? Because, Lovaas wrote, "By being so stingy you can make your rewards work for a long time so that your child will work hard for them for several hours a day."

Imagine being that child.†

Some autistic activists and their allies have compared Lovaas's approach as a version of "gay conversion therapy," the sadistic and misbegotten method to "cure" homosexuality out of people. Indeed, Lovaas himself had personal involvements in projects associated with "gay conversion therapy," so at least he was consistent.[2]

Among the many notable things about *The Me Book*, one stands out in particular. Nowhere in this manual for beating autistic and other neurodivergent children into shape is there a mention of the concept of empathy—instilling it or practicing it. In fact, Lovaas intentionally dehumanized the children he claimed to be helping, writing, "You have a person in the physical sense—they have hair, a nose and a mouth—but they are not people in the psychological sense. One way to look at the job of helping autistic kids is to see it as a matter of constructing a person. You have the raw materials, but you have to build the person."

Is that the mark of a sociopath,‡ to view other humans as not fully human and all human interactions as quid pro quo? It is at least the full embodiment of a dysempath.

WHAT ARE SOCIAL COGNITION AND EMPATHY?

We can see other people in one of four basic ways. We can, of course, humanize them and see them as humans. We can objectify them, viewing them as nothing more than inanimate things without feelings. We can think of them as organic machines, sets of functioning parts, perhaps with

---

* Today, the practice may involve accruing a certain number of tokens that add up to an allowance of a "reward."

† When you do so, it's called perspective-taking.

‡ As in someone whose social behavior is pathological—in which case, it would seem so.

computers for brains. Or we can view them as some kind of animal but not "so human an animal," as René Dubos put it.

These different ways of perceiving other humans impose varying degrees of cognitive burden. Objectifying requires the least effort. It is for this reason that some experts in the art of war emphasize objectifying the enemy. That way, troops can feel that they are simply dispatching unfeeling objects, which limits their cognitive burden and perhaps ensures that the effects of killing people will not linger and cause psychological harm. Viewing people as biological machines requires a little more heavy lifting of the analytical sort: they become a problem to understand and solve. Humanizing others requires empathy: setting aside the problem-solving reflex and feeling their feeling. That's a different kind of burden. And it's one that our DMNs help us bear. A humming default mode network is crucial to the mirroring of others' feelings, perspective-taking, and the expression of human empathy.[3]

Adolescent boys who score high in callous-unemotional traits have reduced connectivity in this network.[4] At the extreme end of a spectrum* are people with antisocial personality disorder. This condition involves intense manifestations of absent empathy and social emotion, insouciance about the feelings of others, and poor impulse control. People diagnosed with it may have abnormally strong connections within the DMN but a disconnect between the DMN and a region in the frontal lobe responsible for supporting attention.[5] One effect of this disruption may be an inability to attend to and recognize emotion and its significance in other people.[6] That might sound familiar to many of us in a fraught political era.

Do you remember from my earlier description of MRI how the magnet aligns the protons, radio waves tug and release them, and the protons resonate back with radio waves of the same frequency? Empathy is kind of like that. When we're aligned in empathy with another person, we feel the tug of their internal state and we resonate with it at the same frequency. Nature built this capacity into us and a lot of other social animals,[7] even though, as you can see from the example of Lovaas, sometimes, nature's work is incomplete.

---

* And not the autism spectrum, as some people think.

In addition to resonating shared emotion, empathy can also involve a cognitive component. That's the part where you're a "mind reader" of what somebody else knows, thinks, or plans to do.* This ability relies on an invisible calculus involving countless factors, from general life experience, to the person or persons themselves, to the specifics of the situation. But broken down broadly, the three features of empathy and this capacity to read minds are (1) detection of emotion that you and the other person are experiencing, (2) taking the perspective of the other person to understand how they're feeling, and (3) sharing the feeling, the resonance, with them.[8] Lovaas failed to do any of these with the autistic children he claimed to help.

To see and be seen through the connectedness of empathy, we have to engage two apparently paradoxical processes: understanding our self separately from the other person while also feeling what the other person feels. Early practice in this strange blend of separate-but-associated occurs when infants and toddlers mimic their caregivers' facial expressions and other motor behaviors. Speech is also a motor behavior that we mimic. This imitation is another kind of resonance between two people, a mirroring of each other.[9] It may be one basis for our empathy and our functioning within a social species.

To achieve this resonance, we rely on our social cognition, the social-thinking capacities of our brains. But you won't find stories of a "social g" or tests developed for the purpose of carving out specific groups as unworthy or less than. Tests do exist, and like all tools they can be poorly forged and badly wielded. But their existence does not trace to a bitter eugenicist root.

As with other kinds of cognition, social cognition is the thinking we do, in this case as we see, process, and respond to what we infer about the mental states of others.[10] It's that invisible math we perform when reading each others' minds. Empathy requires it, and effective social cognition requires empathy. In general, the cognition is the reasoning part of the equation— even if we're unaware of the calculations we're performing—and the empathy arises in large part from sensory and motor sensations with less defined edges and measures. Together, they equal humanity.

---

* A broad phrase intended to capture this capacity is "theory of mind," or ToM, but I don't care for it.

Another difference between global cognition and social cognition is that no one seems to question that social cognition and empathy can be expanded through education. As with global cognition, life itself offers an education in social cognition. It can be rich and deep and broad, filling in all the potential that nature and nurture sketched out. At the other extreme, it can be poor and shallow and narrow, leaving the potential to wither away. Our ability to source learning from other people's behaviors and mimic the social models (other humans) around us[11] allows us to gain skills, even unconsciously. Some people call tapping into our understanding of others' motivation and then mimicking it Method acting. But it's also just a way to build the social-cognition skill set.[12]

Our insight into someone else's story is like reading a narrative of their internal motivations written in real time in their minds. People may have different levels of "mentalizing" ability, but life experience and deliberate practice can hone it, just as practice writing narratives hones the act of storytelling. With storytelling, we resonate with other minds here today, or even the minds of long-dead storytellers.

For all this mentalizing to work, we need some key skills. And we can learn them if we're a little weak in some areas. Empathy, or emotional resonance, is one, along with its component of perspective-taking. We also need to learn to recognize emotions,[13] ours and those of others, imbue our thinking with their influence, and produce output that's genuine, not drama—we have to regulate it. Each of these is a subskill in our social-cognition toolkit. Although some of us start at a higher baseline than others, we all have seeds of these skills, and we can encourage them to flourish in the right environment.

FEELING FOR A LAMP

At the 2020 conference of the Society for Social Neuroscience, Lasana Harris, a social-cognition scientist in the Department of Experimental Psychology at University College London, showed a commercial. In it, a discarded lamp sits, its lampshade bowed, on a sidewalk near some sagging trash bags as the rain pours down. Sad music plays. Above the drooping, discarded lamp, we see a window alight, the replacement lamp setting it aglow.

Then a sardonic man soaked with rain appears on-screen and says, "Many of you feel bad for this lamp. That is because you're crazy." The lamp has no feelings, the drenched man explains, and "the new one is much better."

The next frame shows the IKEA logo.*

The lamp commercial takes advantage of our human inclination to marshal all our social-cognitive and empathic abilities and do two things: adjust them in context, and choose when to use them. As the commercial unfolds, the visuals are deliberately, manipulatively sad: the lamp droops, the rain falls, the setting is a cold, dark evening. Loneliness and longing are both evoked. Contextually, we allow our empathy for this inanimate object to be activated.

And the moment the IKEA guy comes out and yells at us in his Swedish-accented English, a lot of us turn that empathy off.

We weren't having a meeting of the minds with the lamp or resonating with it. But we were resonating with the conditions surrounding the lamp, the implication of abandonment, the isolation, and the feeling of literally being left out in the cold. The lamp's droop recalls the human posture of sadness and loss, a motor expression of a profound emotion. So we take all these cues unconsciously to heart and anthropomorphize the lamp.

Another film is relevant to this inclination, a black-and-white video made in the 1940s that lasts about a minute and a half.† A woman watching it for a study narrated what she was seeing: "OK, so, a rectangle, two triangles, and a small circle. Let's see, the triangle and the circle went inside the rectangle, and then the other triangle went in, and then the triangle and the circle went out and took off, left one triangle there. And then the two [pause] parts of the rectangle made like a [sic] upside-down V, and that was it."[14]

I took notes while an adult I know‡ watched the same movie and commented aloud on it. Here's his narration: "I see a triangle was in a box, and another triangle and a circle came, and another triangle was all up in the little triangle's face, and the circle hid in the box. The circle is cowering in

---

* You can watch for yourself here: https://vimeo.com/291738503.

† The makers of this very short film were Fritz Heider and Marianne Simmel, both psychologists. You can find it on YouTube: https://youtu.be/n9TWwG4SFWQ.

‡ OK, it was my spouse.

the corner—is this supposed to be a family? The circle has escaped! The big triangle is lurking and chasing and trying to find them. The big triangle is mad and breaking things."

Something distinguishes the first commentary on the film from the second: storytelling. The first one contains no discernible story. The second is full of story and its usual ingredients of emotion, conflict, and rising and falling action. It begins with a clash between the big and small triangles, even mentioning a "face" (which is not depicted), shapes "cowering" and "lurking" (the shapes' contours or sizes don't change), the possibility that they could represent a family in conflict, and an emotion word ("mad"). An entire plot emerges of a chase, bullying, hiding, a rescue, a complete escape, and an angry, frustrated tantrum.

What's interesting about this little movie is that it portrays none of the typical communication cues we think of as "human." No faces, no little arms and hands, no body language that even the drooping lamp in the IKEA commercial embodied. Just movement and interactions through movement. The only shape that changes in the entire movie is the rectangle, which is destroyed by the larger triangle. Yet we tend to build a narrative from these elements, in real time, creating it as we go.

The woman who narrated the first commentary didn't do this. It might not surprise you to learn that she had a condition that essentially obliterated the function of her amygdala on both sides. Her cognition was typical, all her senses operated within expected ranges, and her language ability fell within typical parameters. But her narration was limited to only the most concrete, observable facts, without attribution of intention or emotion. That is a powerful illustration of the role of the amygdala in anthropomorphizing, interpretations of movement, and the mental storytelling we engage in about ourselves and others.

Yet the woman with the damaged amygdala could describe and infer motivation using social, emotional, and motion terms when the stimulus presented (a child's book) contained realistic still images of a baby and a dog.* The researchers recounting her case concluded that the child's storybook

---

* From the book *Good Dog, Carl*, which has words on only the first and last of its forty pages.

offered images that were explicitly social, with plenty of context clues. With those clues, she could use her social cognition to find the socially oriented words to describe what she was seeing. In contrast, the movie lacked overt contextual clues having anything to do with sociality, and her amygdala wasn't available to predispose her to see what took place on the screen as a social interaction. Also, interestingly, when the researchers asked her questions about the video and used language that invited a socially oriented reply, she did use social terms.*

This effect of a lost amygdala fits with the idea that several pathways merge at the social-cognition intersection, with contributions from the "top-down" circuits such as our DMN and salience networks, and from the "bottom up," with inputs from the amygdala and insula.[15] We can use unconscious social cognition with an infusion from the amygdala to humanize and construct our stories.

SITUATIONALLY SOCIAL

The DMN and its hubs turn up again and again in studies involving empathy and social-cognition tasks.[16] The authors of an important review have suggested that the DMN sits at the crossroads where the thoughts we generate internally encounter incoming information from the world outside.[17] As we saw with the DMN in the cases involving Cotard syndrome, when these roads do not intersect, disconnects result. The executive-function network (the CEN, and specifically the medial prefrontal cortex) shows up in empathy studies, too,[18] perhaps playing some role in emotional control (maybe it hushes the amygdala if it's being a drama queen) as part of enacting empathy.

Here's where things get weird. Some studies show that early in life, the DMNs of people engaged in ongoing interactions with each other start to *couple*, like a totally normal, real-life, not at all *Star Trek* mind meld. The tendrils of this connection form in infancy, with a resonance developing between a child and a caregiver to create a shared understanding between them.[19] Pairs of students who have close social relationships show neural responses that are more similar to each other's than those of students with

---

* For example, Q: "What was the big triangle like?" A: "A bully."

more distant associations.[20] Even a discussion designed to reach consensus about what a video means can lead to in-sync neural responses when they watch the video again.[21]

A tool still being developed called hyperscanning—the scanning of multiple people's brains at once—indicates that chatting face-to-face causes more neural coupling than if the chatters are sitting back-to-back. Researchers say that this effect implies that nonverbal communication—the body language we use on the ventral, or "window," side of our bodies—is a window into important information.[22]

In similar hyperscanning setups, the neural activity of a person who emerges as an important "node" in a group predicts everyone else's neural activity. Leadership is about mind power, it seems. And like the pairs of students working together on a project, groups of people doing so show greater neural alignment when the project requires them to coordinate and communicate a lot.[23] Even the lowly EEG has been used to show that romantic couples sync when they smile and gaze at each other, to a far greater degree than nonromantic couples.[24]

If this makes you want to go around trying to bend people's minds to yours, please resist the impulse. Motivation alone doesn't foster "brain-brain coupling." Working together toward agreement, consensus, and a common story is also required.

HOW DO WE MEASURE SOCIAL COGNITION?

You might think that measuring something as elusive as empathy or mentalizing (mind reading) would be difficult, especially since we can shut down these capabilities (or some of us can) in certain contexts. But people have certainly tried. Tests have circulated in the research community for decades, and many of them have made it online, where people like you and me can access them.* One of the oldest of these tests, variously known as the Eyes Test or the Reading Eyes Test, is intended to assess mentalizing, or "theory

---

* You can find a comprehensive, open-access list here: www.frontiersin.org/articles/10.3389 /fpsyt.2019.00425/full. An open-access list of the domains of social cognition and associated tests is available here: www.ncbi.nlm.nih.gov/pmc/articles/PMC5928350/table/T1.

of mind." I will explain it, and you can try it for yourself,* and maybe you'll see some of the problems with it.

First, there's a thirteen-page set of instructions.† They list more than ninety emotion-related terms, what the test expects them to mean, and a use of the term in a sample sentence.‡ The test taker is then presented with a series of thirty-six "sets of eyes," each with four of the emotion-related terms listed next to it. They are to choose the word that best describes what the pair of eyes is communicating—that is, "the word you consider most suitable." If the test taker doesn't know the meaning of any of the four words, they are instructed to use the definitions provided in the instructions. The images are old and some of them quite grainy. They look vaguely recognizable, as though they are clipped from pictures of minor celebrities.§

Besides the heavily verbal aspect of the test and the cultural issues (although it has been adapted to non-English-speaking cultures), there's one key issue with it as a putative measure: we don't know the internal states of the people whose eyes we see. The test is supposed to home in on the test taker's cognitive empathy, their ability to recognize signs of a person's internal state and attribute it. But the images are still photos, capturing a split second of someone's entire face in motion, perhaps showing *an* emotion or a fragment of or prelude to one. Attributing an emotional state under those conditions is like seeing the first word of this sentence and thinking you know what the rest of it says.

Facial motion is important to attributing state. A static image simply cannot capture it. If you want an incredible example of how much your mentalizing changes between looking at a still image and viewing one that is moving, find MyHeritage's Deep Nostalgia app, upload a still image—even

---

* The test in various languages and the epic instructions for it are all available on this website: www.autismresearchcentre.com/tests/eyes-test-adult/.

† Some researchers have created versions that are less verbal or that include definitions alongside the words presented with each image.

‡ Example: "REFLECTIVE—contemplative, thoughtful—George was in a reflective mood as he thought about what he'd done with his life."

§ They are actually from photos in magazines, and one of them looks remarkably like George W. Bush, which seems like something that could contaminate outcomes. The word choices with that image are "cautious, insisting, bored, and aghast."

a painting—and gauge the difference in what you attribute to the person in the still image compared to the animated version.

Another theory of mind test is the "hinting task." It consists of ten very short stories intended to avoid putting people with memory or verbal comprehension impairments at a disadvantage (unlike the Eyes Test). Each story has only two characters and involves a quick exchange between them. The second character always says something a little oblique instead of just coming out and stating a need or desire. The indirect statement offers a hint—hence the name of the task—to the test taker about what the second character really wants.* A test administrator asks the test taker to describe what the second character actually wants. If the answer is right on the first try, the test taker gets two points. If it's still not right following a hint and a second try, the test taker gets zero points for that question. The maximum number of points is twenty.† Perfect stores are not uncommon.[25] An example: "Rebecca's birthday is approaching. She says to her Dad, 'I love animals, especially dogs.' What does Rebecca really mean when she says this? What does Rebecca want her dad to do?"‡

These two tests fall into the "reading-the-mind" category of tests for empathy. There are other batteries designed to serve this purpose, but they tend to be proprietary, usually require a trained administrator, and are not freely available online. If you're interested in testing your general empathic capacities, you can try the Interpersonal Reactivity Test, which is available for no charge at https://mysocialbrain.org/iri_info.html. It is described as encompassing four subscales: perspective-taking, empathic concern for the feelings of others, distress at the pain of others, and being able to immerse oneself into fantasy.§ Taking the test entails agreeing to your responses being used as part of a study at Dartmouth College.

---

* It really comes across as an exercise in how not to ask directly for what you want.

† You can find a version of it, translated from Spanish, at www.elsevier.es/en-revista -revista-psiquiatria-salud-mental-486-articulo-adaptation-hinting-task-theory-mind -S2173505012000325.

‡ The test taker is supposed to say that Rebecca wants her dad to get her a dog for her birthday.

§ The theme of imagination and stories is one that will come up again in this chapter.

The test is quick . . . I took it myself.* As far as what to do with the information, the survey gurus say, "Only you can decide how to appropriately weigh the information the survey provides." If you find that your scores aren't in a window that feels comfortable to you, the rest of this chapter looks at ways to exercise your social-cognition muscles. But as the Dartmouth researchers note, "Receiving a high or low score on this survey is not necessarily a good or bad thing. There are both advantages and drawbacks to being very empathic."

Our social relationships matter to our overall health. A 2010 meta-analysis of 148 studies showed that people with strong social relationships have 50 percent improved odds of survival.[26] To translate that into language that doesn't make it sound like those folks might be immortal, what it really means is that within a given time frame, 50 percent more people with strong social relationships will be alive compared with those who have weaker ones. The effect was stronger than if someone with heart disease quit smoking. It was even stronger than the effect of abstinence from alcohol compared with drinking more than six drinks a day.

The power of this effect has led to weak social support being viewed as a "fundamental cause of disease." If you are wondering what your social support is like, well, there's a scale to measure that too: the Multidimensional Scale of Perceived Social Support.† It consists of three subscales covering a significant other, family, and friends. To score it, you add up the responses within each subscale and divide them by four to get the average of the four values. Once you have those, add them up and divide the total by twelve for your scale score. Average scores in different populations‡ range from about 5.8 to 6 (out of 7) but depend on culture.

---

* I scored higher for overall empathy than 82 percent of other test takers, higher for perspective-taking than 89 percent of other test takers, higher for empathic concern than 88 percent, and higher for personal distress at others' pain than only 29 percent, which I attribute to having had to keep my cool through dozens of emergencies as the parent of three children. For the fantasy subscale, I scored higher than 89 percent of other test takers.

† You can find the scale available for free at www.tnaap.org/documents/mspss-multi dimensional-scale-of-perceived-social.pdf. This website doesn't offer an explanation of how to score the scale, so I've included that information in the text.

‡ College students in the United States, pregnant women, European adolescents, and doctors early in their training, among others.

The upshot is that the healthier our social networks are, the healthier we are, and not just because we are in a hiking club or a vegan-cooking clique. When we are socially isolated, our risk for mortality within a certain span of time increases.

HOW CAN WE TAILOR SOCIAL COGNITION?

We know that we can enhance social cognition and empathy because none of us are born with either fully formed. As with global cognition, the power and the potential to increase them are in place—lots of neurons, lots of flexibility to work with. But realizing the prospects of both requires human contact and, yes, learning from other humans. We may start somewhere on a continuum of inborn instinct for social cognition, but the knowledge begins at zero.

For social cognition compared to general cognition, there's a lot less controversy over the idea that this capacity can be built on existing potential. And there's a lot less hand-wringing and genome-wide association investigations of whether certain subpopulations vary genetically in their perceived capacities for social cognition. I can't think why that might be.

We've already seen how we begin this learning by mimicking the people we spend the most time with in early life. These are our first experiences of resonance. Mimicry is one of the two ways we update our social-learning knowledge base.[27] In addition to discovering the power we can wield with our first toothless smiles, we watch other people to figure out what works in different environments and contexts. In this way, we ascertain what's socially relevant in those environments, increasing their salience.

When we struggle with this learning, encounter life situations that inhibit it, or turn it off as a protective measure, we can lose those skills . . . for better or for worse. What, exactly, can break the learning process?

WHAT MAKES A DYSEMPATH?

The DMN initially gained a reputation for being the network that's engaged when we're daydreaming or when we're not performing a "goal-directed task," such as remembering a string of numbers until someone asks us to repeat them. As studies about this mystery network began to emerge, new information suggested that it wasn't just a "task-off" structure but more like the orchestra leader. It is always busy, sucking up most of the energy our brain uses, and setting the pace for the other networks. But one thing that can really muffle the DMN and leave us "screaming inside our hearts"* for relief is cognitive overload.

When a task is routine, you don't devote a lot of attention to it, and your DMN can run on a higher setting. You can be cleaning the bathroom or folding towels or taking your daily walk, and as long as those are everyday activities that you don't have to map out every two seconds, your brain is free to fire away from its armory of self-generated thoughts. But an extra cognitive load—such as worrying about whether your teenager is wearing a face mask at school or how well you'll weather dose two of a vaccine—might interfere with your ability to automatically complete a task, causing you to pause and remember what you need to do next, breaking your reverie.[28]

Oh, also, when we're cognitively overloaded, and our DMNs are deranged, our social cognition may be dampened as well.[29] You know what happens when we lose the ability to navigate socially? We stop understanding the other brains in our brain collectives. We become rigid and black-and-white in our thinking because doing so lessens the load. In other words, when the information tsunami swamps our minds, it washes away our capacity for empathy, understanding others' emotional states, and cooperative problem-solving.

The DMN, as noted, uses far more than what seems like its fair share of the energy allocated to the brain, and the brain in turn uses more energy than its fair share of what we take in. For some reason, this skewed allocation developed and persisted in human evolution, implying that there is something crucial about using our DMNs and having brains tailored for social cognition and understanding one another. As the authors of one study asked, does the fact that our DMN hums along during down time leave us primed to "conceive others as minds and not merely bodies?"[30]

---

* This was what Japanese authorities suggested to people riding roller coasters during the COVID-19 pandemic—rather than really screaming and broadcasting potentially virus-laden spittle all over everyone.

Lots of books have been published with lists of ways to become more empathic. I've included a few adapted checklists at the end of this chapter. But you may have picked up on the concept that our ability both to appreciate storytelling and to tell stories ourselves has a role in building empathy. We can't write about people who are not us—especially people we just make up—if we can't get into their real or fictional minds and figure out what makes them tick. So perspective-taking is a prerequisite for storytelling.[31]

And it's required for those reading the story, too, in complicated ways. The reader (or listener, or viewer) must have some insight into the intent of the story's creator, peering between the lines, drawing inferences and making connections. This is all excellent practice for understanding other people because these *are* other people. If we've learned anything so far about what we can do with the brain, it's that practicing a skill makes us better at that skill.

The reverse is that as the audience for stories, we do the same: we care about the total stranger on the screen pretending to be someone else (aka an actor) because we put ourselves in the narrative, in that person's place, feeling their sadness, fear, anger, happiness, or surprise. The creation of art through storytelling and the meeting of completely separate minds at its instantiation is a truly human attribute. It's what allows us to read about men who died of brain injuries in Egypt five thousand years ago and feel their agony. It's why Dickens's *A Christmas Carol* is re-created again and again, allowing us to share the yearning for redemption and a happy ending during a season of giving and forgiveness.

When you read lists of steps to improve empathy, you'll find that a common theme is the twin capacity to be open to new experiences and to seek out stories and storytellers. Philosopher and author Roman Krznaric, writing in 2012 about the "six habits of highly empathic people,"* gives these habits as follows: "Cultivate curiosity about strangers. Challenge prejudices and discover commonalities. Try another person's life. Listen hard and open up. Inspire mass action and social change. Develop an ambitious imagination."[32]

---

* In keeping with the usual "what's in it for me" mode of popularized brain-related offerings, he notes that "empathy is not just a way to extend the boundaries of your moral universe. According to new research, it's a habit we can cultivate to improve the quality of our own lives."

You'll note that the threads weaving through the six edicts relate to listening to and telling stories, and being open to new experiences and change. Every time we reach for those qualities, we add to our repertoire of understanding and expand the frontiers of our social cognition. If you recall Raymond Cattell's study of rural versus urban families, he found higher intelligence scores among the families who moved to urban areas. His conclusion was that people who move away from the confines of rural life are more intelligent. But the greater exposure to change, varied stories, and new information typically encountered in a more populous place may well have been the factor in play.

Or perhaps the families who relocated were more open to new experiences. Openness leads us to seek out stories. They expand our cognition, whether we tell them or have them told to us. Both lead to change, and change translates in the brain to experience, memory, and feeling. This experience, in turn, can lead to broader cognitive effects, because what are we doing when we listen to stories if not constantly taking the perspective of the characters and working out how they might solve their problems?

Even when the story ends, some of us cast about for alternate endings, ways that the outcome could have gone differently, that one moment when another choice could have shifted fate. This mental time travel is something we have to do in real life as we visit and revisit our experiences and draw lessons from them.[33] These are the processes of problem-solving, and problem-solving is the key feature of what we view as global cognition. In this way, we help others help us. And we achieve that treasured goal: far transfer to more global cognitive powers.

Radivoy Kvashchev, who conducted the Belgrade study, had some intuitive sense about these connections. After all, the way he got the attention of high school students and focused them on problem-solving was by beginning with wacky yarns, such as an improbable problem involving pirates and a man in a terrible plight unless he could get some bread into his mouth. The students not only problem-solved during these exercises. They had to put their heads—and the brains in them—together to do it.

To put a little bow on this tidy gift, some people have called the brain network common to these skills the "imagination network." Indeed, some researchers have posited that this network, known to you as the DMN, over-laps these cognitions and capacities because of its role in distinguishing self from other, mentalizing another's thoughts, daydreaming or fantasizing (off-task spacing out, which is key to imagination), episodic memory, and sorting out alternatives.[34]

Aristotle noted in 330 BCE that "man tends most towards representation and learns his first lessons through representation."[35] Developing this understanding through fiction relies on the same processes we use to understand the real-world stories around us. Lest you think I'm just telling stories, research shows that people who read a lot of fiction do better on tests of empathy than people who stick with nonfiction (that presumably does not tell stories). A 2018 meta-analysis found a small but real improvement in social cognition among people who read fiction compared to those who don't read much at all or who read only nonfiction.[*] As the authors wrote, "Open a work of literary fiction and you immediately gain access to the inner workings of another person's mind."[36] The effect they found, though small, was one that they argued "has the potential to be very meaningful." Interestingly, most of the studies they included in their analysis had participants read only a single fictional short story. The sole study that required participants to read a complete fiction book also showed the largest effects on social cognition.

Another study using functional imaging to monitor the brains of people reading fiction found that specific parts of the DMN showed a response to "especially vivid" content, and other parts responded to passages with social or abstract content. Activity in the latter part of the DMN[†] was associated in turn with the readers' best performance on tests of social cognition. The authors of this study drew a direct line connecting reading fiction to activity in this DMN region to better performance on the social-cognition tests.[37]

---

* But please don't stop reading this nonfiction book, because there is still an effect of nonfiction reading on empathy and mentalizing—it's just not as profound as the effect of reading fiction. Plus, I tell some stories!

† The dorsomedial prefrontal cortex.

How long any of these effects lasts is unclear, but the good news is that reading or listening to stories usually can be a lifelong,* even daily activity. You can expand the effect, possibly, if you choose fiction that is about unfamiliar people, places, and things, showing openness along with embracing story. Watching movies may have a similar salutary effect on aspects of social cognition, and so might playing video games that have a narrative component.[38]

THE MINDFULNESS-EMPATHY CONNECTION

A key aspect of mindfulness is to approach the experience without judgment. To practice empathy, you have to do the same thing: approach the person (or character) without judgment, and edge your way into their perspective—by asking yourself not "How would I feel?" but rather "How would someone who is this person, with their life experiences and in their situation, feel?" And then show them that you understand the feeling. It's important at this step not to take over the experience for yourself—you're resonating with the person, not trying to dominate the frequency.

Mindfulness practice itself has been associated with better emotional connections, empathy, and some social-cognition measures.[39] A study that matched thirty nonmeditators with thirty meditators found that the latter group scored higher on empathy, emotion recognition, and mentalizing (reading others' minds). They interestingly did not score higher than nonmeditators on a personal-distress subscale of empathy, and I think I know why. The questions tend to be about reactions in crisis situations, in which people are in distress, and if you can keep calm, you're going to score lower on such a scale. A study of medical students in Italy found that those with more mindfulness-type inclinations tended to experience less emotional distress amidst social tension.[40]

A 2018 meta-analysis of randomized controlled trials of the effects of meditation on positive social behaviors also showed small to medium effects.[41]

---

* And studies certainly indicate an association between early exposure to stories in books and children's social cognition, especially if parents also make space for discussions about the characters' inner states.

The comparison groups tended to be passive—they received no treatment at all or were wait-listed—and outcome measures were self-reported, which could bias the findings. But with a medium effect, there's probably some there there. One reason may be that the forms of meditation most used in these trials were compassion meditation and loving-kindness meditation, which deliberately target developing empathy and prosocial behaviors. The specific aim of loving-kindness meditation is to increase compassion, loving-kindness (natch), empathic joy, and equanimity. Practitioners seek to develop awareness of feeling kindness and compassion for others, using visualizations or phrases or other tools that keep their minds on this positive orientation. Compassion meditation focuses specifically on walking through the experience of someone else's suffering and how to soothe that suffering.

Several other investigations have similarly linked the practice of meditation generally to improved social cognition and empathy, including at least one 2020 study with active controls.[42] The authors reported that breaking down daily contemplative practices into three-month modules that they had developed—specifically, a presence module, an affect module, and a perspective module—targeted specific aspects of social cognition. They found the greatest effect on attention with use of the presence module, on compassion with the affect (emotion-recognition) module, and on mentalizing with the perspective module.

They also found that the sequence in which people participated in the modules over the course of nine months seemed to matter. The only module that affected mentalizing scores was the perspective-taking practice. But "stabilizing the mind" with the presence module was associated with bigger effects than the affect module had, linking attention or salience (and the salience network) to the feeling of caring about someone else. It's true: you have to notice that they need the care before you can care. And these researchers found that although the affect module, as predicted, was associated with increased compassion scores after the intervention, the perspective module also led to increased compassion scores. They concluded that the findings point to two ways to reach compassion: practicing emotion recognition and practicing perspective-taking.

Indeed, the emphasis on *practice* may be key. (It is also a theme of this book.) Improvements in empathy have been seen from practicing self-reflection on social experiences with a focus on being nonjudgmental.[43] That makes sense because empathy itself happens when you shut judgment down.

The saying is that "practice makes perfect," but what if practice simply makes preparation? Did you know that you don't have to wait for conflicts or tensions to arise before figuring out how to deal with them? With some practice of key aspects of a process called "wise reasoning," you might be able to preempt interpersonal dramas. Wise reasoning in the face of conflict requires elements similar to the ones people use in mindfulness practices and empathy, including:

- a lack of judgment (having intellectual humility)
- openness to new situations (recognition that things are uncertain and can change)
- perspective-taking (perspective-taking)

When the tension and the anger build, hit a mental pause button. Go through the three elements of wise reasoning and assess whether or not you're applying them. This review could leave you feeling OK after a stressful interaction, rather than seething with tethered resentment.[44]

Finding ways to avoid seething can be important these days. Even before this country and our families and friends became divided by an ever-widening socio-political chasm, we in North America dealt with an average of *seven* conflicts a day in our social circles. It's not possible, of course, to map out or anticipate every conflict. But we can acknowledge the fracture lines and attempt to stitch them up with a little care and planning, using all the cognitive tools in our tailoring kits . . . along with a dash of wisdom. Wisdom, which has been defined most succinctly as applying sound judgment in difficult situations, encompasses the three social-cognition elements of wise reasoning. It also includes a willingness to seek compromise—not to cave in to another's demands, but to meet them halfway, if that's your goal.

A focus on understanding others—not condoning what they do, but comprehending it—could improve our own happiness. Brain imaging has suggested some links between self-reported happiness and a DMN network that is anticorrelated with the salience network and in communication with the central executive network. High scores for "trait empathy" (being consistently empathic rather than using empathy situationally) are associated with similar patterns.[45] The implication is that your sense of happiness and being empathic overlap in the brain's connections. It's just an implication, but if a feeling of happiness is a side effect of strengthening your empathy, that seems to place another finger on the scale for empathy.

Our brains are senders *and* receivers. What we've variously called "vibes" or "extrasensory perception" or "mind melding"—that sudden chemistry we feel with another person—has a very real biological basis.[46] It developed right along with our species over millennia and was reinforced in the hubs within our brains, and by our brains acting, reacting, and interacting with the other brains around us. Our ancestors got us to this point, using these features to prop up *H. sapiens* in an ongoing communal existence, across time, cultures, and space.

In this way, we often find the middle ground—whether through natural or learned empathy, wise reasoning, or sheer luck—that leads to a shared understanding and the uncovering of commonalities. Through art, especially the art of storytelling, we perpetuate that understanding and undergird those commonalities. With stories, our ancestors reach out to us and other cultures speak to us, expanding that rapport to include millions of humans we will never even meet.

Right now, we are the ancestors of future generations. We can leverage the flexibility of our social cognition and empathic skills, and deliberately, wisely optimize them for our own good and that of others, including those who will come after us. Or we can choose not to, dehumanize one another instead, and let the world burn.

CHECKLISTS

Brené Brown's Empathy Skill Set

Brené Brown has discussed empathy as a set of skills, the tools in your tool-kit for tailoring your own empathy.[47] The list, according to Brown:

- ✓ Stay out of judgment.
- ✓ Take someone else's perspective.
- ✓ Walk in another person's shoes.*
- ✓ Communicate what you understand about what the other person is experiencing.

Six Habits, Modified

Another well-known checklist is the "six habits of empathic people," developed by philosopher Roman Krznaric (and mentioned earlier in the chapter). Below is a modified version.

- ✓ Grow your curiosity, including by talking to new people—possibly from six feet away and behind a mask, but still.
- ✓ Examine your biases,† and look for things you share with someone else.
- ✓ Get near the real-world experiences of others (the actual advice is to "try another person's life," but that seems like privilege cosplay to me), perhaps by volunteering or finding ways to serve those who are underserved.
- ✓ Listen and be open in mind and spirit.

---

* You can try this out in the literal sense with a tour through an exhibit called *A Mile in My Shoes* at the Empathy Museum. (Yes, there is an empathy museum. It's a traveling series of art installations.) Visitors can put on a pair of someone else's shoes while listening to that person narrate their personal experiences.

† You can take a test of your implicit bias here: https://implicit.harvard.edu/implicit/take atest.html.

✓ Leverage your privilege for social change (Krznaric calls for "mass action and social change," which would be great, but it's a bit of a big ask, I think).

✓ Deliberately nurture your imagination and give it wings (the original says to "develop an ambitious imagination," which is lovely, but nebulous in the adjective "ambitious").

Wise Reasoning: Five Things

Practice wise reasoning, especially for interpersonal tensions and conflicts.[48] Below are five key aspects to keep in mind:

✓ *Know the limits of your own knowledge.* This understanding puts us in our place, contextualizes us individually, and helps us approach interpersonal situations with humility.

✓ *Use that global cognition of yours with social cognition to play out different causal chain-of-event scenarios to predict the different directions a situation could go.*

✓ *Practice perspective-taking.* Almost any engagement with another human being (or any living thing, in my opinion) should involve this practice. Don't forget: practice builds the skill.

✓ *Practice meta-perspective-taking.* Look at the situation from the outside in. How would it appear to someone who's not intimately involved? Would it have the same emotional valence, intensity, relevance? Sometimes stepping outside a situation and observing it can help you see a resolution without the drama.

✓ *Cast around for the middle ground.* Where that is and what the negotiating terms are will be highly personal. Finding a happy medium may not even mean that a resolution to the problem is at hand, but it can leave you with a feeling of meaning from the experience.

CHAPTER 6

STRESS AND ANXIETY COGNITION: LIGHTENING THE LOAD

ANXIETY HAS A long and storied history, at least in the Western canon. Early philosophers had ideas for ways to achieve a calm state of mind, implying somewhat erroneously that the noncalm version was not particularly desirable. Some version of panic, phobia, or posttraumatic stress is even present in texts attributed to the Greek physician Hippocrates, which include the story of a man, Nicanor, who feared the flute girl so much that he'd have a panic attack when he heard the sound of a flute. But only at night, rather than during the day. People seemed to recognize phobias as a thing, and Cicero wrote of anxiety as a disorder, even going so far as to distinguish an anxious disposition (trait anxiety) from acute anxiety (state anxiety). Cicero and his Stoic philosopher brethren wrote of exercises that might clear the chaos and restore tranquility to a stormy mind. From there, we can ask the question what is the difference between anxiety and stress, at least as we view and define them today?

Judith Moskowitz, a professor at Northwestern University and an expert on stress and positivity, told me that stress arises from our relationship with our environments and how we appraise a situation: Is it taxing? Does it exceed our resources or endanger our well-being?

Modern practitioners define stress in two phases. The first is *cognitive appraisal*, the determination you make about what an interaction with your environment is doing to you. The second is *coping*, how you manage it or don't manage it, if your appraisal leads you to conclude that the stress has chewed up all your resources for coping.

Anxiety can be a completely normal and utterly useful state that drives us to gnaw at a problem and try to predict outcomes. Like making a to-do list, creating a list of outcomes can have a paradoxically soothing effect. As with stress, though, when anxiety becomes chronic, that is a problem.

Diagnosable conditions related to stress or anxiety include posttraumatic stress disorder (PTSD), generalized anxiety disorder (GAD), obsessive-compulsive disorder (OCD), and social anxiety disorder.[1] This book does not make any recommendations regarding diagnosed conditions except to urge you to seek professional insight if you are concerned that you have one. It does summarize studies of interventions for stress or anxiety, some of which have focused on people with these conditions, as an example of what a particular intervention might do. The marker of needing help has typically been if you find symptoms interfering with your daily activities, relationships, and success.

Chapters in this book have focused on global cognition and social cognition, categories of thinking that overlap. In this chapter, we're going to look at what I call stress cognition: how we recognize and think about stress, and ways to channel that thinking so stress doesn't leave us in a constant state of teeth-grinding anxiety. The good news is that our stress cognition can be sharpened, just like our skills in other realms of cognition.

A STRESSED NETWORK

May 12, 2008, was a Monday. That morning, at around six thirty a.m., an 8.0* magnitude earthquake hit Sichuan Province on the eastern edge of the

---

* Some sources put it at 7.9.

Tibetan plateau where western China's iconic mist-wrapped mountains rise from the golden expanse of the tablelands. The quake resulted from "excessive rock fracturing," possibly from water pressure building up behind a 150-meter dam erected in 2004, redirecting enough flow into surrounding fissures along a fault line to trigger the temblor.[2] It shifted the Earth's surface in some areas by many meters.[3]

The town of Yingxiu, which lies tucked into the green embrace of rocky ridges, was the quake's epicenter, where 5,462 residents died and most structures were destroyed. Overall, more than 100,000 people—thousands of them schoolchildren—died from the quake and the disasters that followed it, which included hundreds of thousands of landslides and avalanches.[4] It was by any measure an epic disaster that also proved to be chronic, with aftershocks occurring for months afterward and landslides triggered by fresh rain, the water turning the earth into liquid.

In the aftermath, parents of children who had died in collapsed classrooms, including in Yingxiu, joined together in a growing chorus, leveling corruption charges against those responsible for constructing the schools. Under threat of government surveillance, these parents, along with artist and activist Ai Weiwei, spoke out against what they saw as a government turning away from the reality of this corruption and the resulting loss.

One parent whose twelve-year-old daughter had died in a destroyed school told the Associated Press that she wanted the government to ensure that the people who'd defaulted on their duties were seriously punished.[5] Her two trips to petition the powers in Beijing had turned the eyes of the central government on her, but she was undeterred. "I hope the government will really do what they say they would and not brush off us parents," she said. "If this is the case, the hearts of my husband and I will be more at ease." A few months after their loss, she and her husband found that they were expecting another child. She told the *New York Times* that she was working to see the pregnancy as "the return of our daughter."

Another father who lost a son in the same school collapse told the *New York Times* that the birth of a son within a year of the quake did nothing to assuage the urgent desire for justice he and his wife felt.[6] "If we can't get justice, we'll have our son carry on the quest for justice. This issue will be a burden on this child." Also placed under surveillance, he said that he had been threatened with imprisonment if he kept up his protests.

These bereaved parents were engaging in a form of reappraisal, seeking a way to recast their thinking and actions in response to the horrific loss of their children. In keeping with reappraisal practices, in this case likely unconsciously, they sought to use an adaptive approach oriented toward altruism—helping others and ensuring justice for those who died and those still in danger due to shoddy construction—to gain some kind of meaning from an unimaginable yet all too real tragedy. A group of parents who filed lawsuits included a "public apology" as part of their request, but the lawsuit was dismissed. Unfortunately, their quest for justice—and to prevent future injustices—led to a government crackdown on the families who were protesting what critics in 2011 called "one of the worst abuses of human rights in recent years."[7]

With so many stressors burdening the survivors of the devastating earthquake and its aftermath, you might think that their brains would show similar signs of the effects. To find out, researchers from several universities in China used fMRI to assess the connections among major brain hubs in earthquake survivors who had been diagnosed with PTSD, and compared them with those of survivors who did not meet PTSD criteria. The exams, performed within nine to sixteen months of the quake, measured sixty-two people with PTSD and sixty-two without it, matched for age, sex, and education level.[8] Outwardly, the groups differed in their stress responses to an enormous and ongoing disaster. Did they also differ inwardly?

They did. The DMNs of the survivors with PTSD differed in their connections and connection strengths from the DMNs of survivors without PTSD. The people with PTSD had weakened connections between the DMN and other brain regions, and increased activity related to the hippocampus, which gates memory. The implication was that a memory gate had been left open, serving up painful scenes that drew the attention of the DMN away from its usual involvements. The findings for the DMN disruptions in PTSD were not new.[9] In soldiers who have experienced combat and have PTSD, the worse their symptoms are, the more disconnected their DMN function.

You may recall that the salience network tells us what is important around us. When it's operating in a healthy way, it does so selectively. When we space out, it quiets down and lets the DMN take the wheel. But a hyperactive salience network is always on the lookout for danger; in a pattern identified in people with PTSD, it doesn't dial back when the DMN is active. The hippocampus may keep the two networks inappropriately amped up and wary, serving up alarming reminders of memories and keeping the brain in a state of vigilance.[10]

What remains unknown is which came first. Was there already a vulnerable DMN that, once disconnected, stays that way, leading to worse PTSD symptoms? Or does the severity of the experience disconnect the DMN so powerfully that unaided repair seems impossible?

In people with anxiety disorders, the DMN and the central executive network rev together, rather than alternating dominance as they would under typical circumstances.[11] They also rev together during creativity-associated tasks, suggesting that our capacity for creativity might feed into our anxiety, allowing us to catastrophize all kinds of terrible scenarios. The amygdala is also very likely weighing in a lot more than usual in states of anxiety.[12]

The DMN in someone with anxiety shows atypical connectivity compared to the DMN in someone without it.[13] Anxiety disorders in general involve an overbearing but atypically connected DMN because the brain is always on task and on guard. The result of being alert to potential threat from every sound or movement is that little time remains for self-examination, self-positioning in relation to others around us, and, most crucially, replaying autobiographical experiences and making sense of them. These difficulties mean that people with anxiety can have a hard time making sense of self. Some who have this sense of disconnection from self, which may be quite real in the brain, express feeling like an object rather than a person. Recall from the social-cognition chapter what happens to empathy when we objectify other people: we turn it off. People with anxiety may be turning off empathy for their own sense of self by objectifying it.

The reverse is also true. One research group assessed what happened to the brain when a person was physically touching another person (holding their hand) during an unpleasant experience (in this case, a quick electric shock). They found that the dorsal anterior cingulate cortex, which sounds the alarm about danger, showed less activation during a hand-holding shock compared to a shock with no hand-holding. I'd like to think that's why people clasp hands or hug each other in times of duress or threat, but the results suggested an effect only when holding hands with a partner or friend, not with a stranger.[14] Perhaps with a more intense threat, a stranger would do.

The dorsal anterior cingulate cortex and insula, both part of the salience network, are associated with how we respond to social stress. These regions have even been linked to bodily inflammation in response to the social stressor of being rejected[15] (yet another reason to evaluate your social networks and perhaps loosen connections with anyone who leaves you with a feeling of rejection). They work with some cortical hubs of the DMN and their connections to the amygdala (emotion), hippocampus (memory), and thalamus (everything) to control important bodily functions, including heart rate and the immune system's response to stress. Even blood pressure increases from the stress associated with changes in some of this connectivity.

Brain is body, and body is brain, and they are both keeping score. That's reason enough to take steps to alleviate the burden and harm of excessive anxiety and stress.

HOW CAN WE TAILOR OUR STRESS COGNITION?

In my discussion with Judith Moskowitz, she described broad categories of coping mechanisms and emphasized one set of tools in particular. If the thing that's stressing you is a problem you can solve, focus on the situation and try to resolve it. Doing so, she said, is being a "problem-focused coper" and is adaptive (in a positive way). It also happens to call on the core activity of global cognition, solving a problem, which may draw energy and attention from your stress.

By contrast, running away from a problem (my words, not hers) that you could sort out is probably maladaptive because this avoidance (a) keeps you from getting comfortable with the process of coping and (b) leaves the problem unsolved. Make it a habit, and your life will become a trail littered with messes that won't disappear.

The key is being able to identify whether a difficulty is solvable. Echoing Reinhold Niebuhr's oft-quoted serenity prayer,* Moskowitz said that if you can truly do nothing about a stressor, then avoiding it or distancing from it is adaptive in a positive way. She cited the COVID-19 pandemic as an (obvious) example of when avoidance is good: "If I did something to take my mind off of the problem, that is incredibly adaptive in this situation"—and better than thinking about the crisis twenty-four hours a day.† Some behaviors, she pointed out, are always maladaptive, such as excessive drinking. But aside from such exceptions, whether something is adaptive or maladaptive is situational.

What does adaptive coping look like? Moskowitz points to another devastating pandemic: the HIV/AIDS crisis that haunted the 1980s. She worked on a research study involving men who were caring for their partners with AIDS and seeking methods to cope with the stress, including the strain of caring for someone dying because of a virus that they also might have. "There are few situations more stressful," she said. But early in the study, one thing cropped up repeatedly in her discussions with the men. They kept saying, "You're not asking us about the good things that are going on," Moskowitz recalled. "That was surprising . . . *let us talk about the good things.*" When the men did, they spoke of positive things that got them through the day. "In almost every single interview, they could come up with something," she said. "Often it was really small; the sunset was really beautiful, a walk on the beach, a beautiful flower, 'making a meal for my partner that he appreciated.'" These little notes of beauty sound a lot like brief pauses to cultivate mindfulness.

Moskowitz remembered that she and her colleagues were focusing on the stress of bereavement, and meanwhile, the study participants were focusing on things outside that realm of experience. The realization prompted her to study how positive emotion in the midst of stress affects health outcomes. "How is that helpful?" she asked about this positive emotion, along with "can we teach people these skills, frame them as life coping skills?"

---

* "Grant me the serenity to accept the things I cannot change, the courage to change the things I can, and the wisdom to know the difference."

† I don't know about you, but the pandemic managed to invade my dreams, so twenty-four hours isn't necessarily an exaggeration.

She has found that the answers to these questions are "in many ways" and "yes." In a series of studies "named after plants and flowers,"* Moskowitz and her colleagues sought to teach a package of skills to study participants. The IRISS study included participants who had recently learned that they had tested positive for HIV. Those who were taught coping skills had some unexpected outcomes compared to their counterparts who received usual supportive care. Yes, they had more positive emotion, which was an aim of the intervention. But they also had a greater likelihood of a reduced viral load and were less likely to be on antidepressants. Control participants experienced an increased average use of antidepressants.[16]

The researchers plucked skills for the coping program from the literature, and included in it noticing positive events, gratitude, mindful awareness, and positive reappraisal (discussed below). Moskowitz and her colleagues have tailored them for different groups, including people who care for loved ones with dementia and people who have type 2 diabetes or depression.

Moskowitz is careful to draw a distinction between mindful awareness of positive things amid impending loss, and the belief held by some that if you aren't constantly cheerful, your ill health is your fault. "That's sort of the toxic positive thing," Moskowitz said about the latter. "Just put on a happy face and you'll live longer. This is not that." The "toxic positivity" message is a terrible one, she added. Rather than pushing positivity on people, she said, "It's important to just say, 'That really sucks, I'm sorry.'"

What do she and her colleagues think about the lower viral loads in people with HIV who completed the program, if the takeaway isn't "put on a happy face and your health will be better"? She said, "In terms of 'blaming the victim,' we are really careful to say that life is hard, what you're going through is hard. It's normal to feel negative emotions." Their program doesn't urge participants to ignore or try to avoid feeling their negative emotions. "It's possible to feel positive emotions alongside the negative emotions," she noted. One hypothesis for why viral loads might be lower for people in the program is that those who are in touch with flares of positive

---

* IRISS, ORCHID, LEAF, LARKSPUR, and so on.

emotion might be more likely to eat well or follow clinical recommendations. But, said Moskowitz, "We haven't done the study where we connect all of those dots."

Most of the examples of precious positive moments that the men in the AIDS study cited involved nature: a walk on the beach, a sunset, a flower. Even small doses of nature can offer a bit of relief and feelings of awe and appreciation when they're otherwise hard to come by. Somehow or another, whether you befriend a special plant and visit it every day, let the sun shine on your face through a window or outdoors, or dig deep with a miles-long hike, foot to ground in a soothing rhythm . . . most of us can make time to come into contact with the natural world. Research indicates that communing with nature is linked to reduced mental distress, better memory and mood, and, for people with PTSD, a reduction in symptoms.[17] It is restorative to body and spirit because we, too, are nature.

You don't have to carve out hours each day. You've probably heard about how bursts of high-intensity exercise can offer benefits that match or improve on those afforded by a miles-long run. Interacting with nature is similarly beneficial: even just a few minutes a day spent in nature contact has been associated with a better sense of well-being and reduced stress and anxiety.[18] Doesn't that sound nice?

We are, as mentioned, part of nature, no matter how many walls and layers of clothing we try to erect between ourselves and the natural world. So I speculate that it is possible that those ten minutes or so in communion with the outdoors could synergize with the effect of having another human around, one with whom we resonate naturally. A friend, the fresh air—and a fizzing away of your stress.

TRIPPING THROUGH THE TULIPS

I'll mostly leave discussion of prescription drugs to the specialists, but I will say that they don't help all or even most people with anxiety disorders. These are findings averaged in clinical populations, so that doesn't mean they won't work for any given individual.[19] You should, of course, meet the criteria for any condition before being prescribed a medication for it.

But let's talk about psychedelics, which some practitioners swear work better when mixed with a high dose of nature. They obviously can't be prescribed, because they are completely illegal. So I'm in no way urging anyone to try them. The people who ran a "'self-blinding' citizen science initiative"* were a little more invested than I. They recruited 191 people to a placebo-controlled trial in which the participants arranged for their own placebo and "intervention," that is, their psychedelics, to assess effect on outcomes including mood and creativity. They were given online instructions for incorporating "placebo control into their microdosing routine"†—which is a practice I do not observe, but apparently some people do.[20]

The trial used a design that seems to be gaining traction: N = 1. "N" is the number of participants in a trial, which, as the = sign suggests, was 1 in this case. But hark, you say, perhaps raising a hand: I just told you there were 191 people, not 1. Let me explain. Each participant in an N = 1 trial partakes of both the intervention and the placebo, in a sequence, back and forth a few times, without knowing which is which (that's the self-blinded part). When 191 people do that, you end up with a placebo-controlled trial with 191 participants, each person their own little self-trial.

Participants had to set up their own placebo, which you'd suspect would introduce a bit of confounding, or at least a temptation to peek around the blinders. The instructions said to prep two sets of capsules, one containing their microdose drug of choice (mostly LSD or psilocybin) and the other containing nothing. To hide the "nothing," participants had to use non-transparent capsules for both preparations.

After preparation, each capsule was placed in a separate baggie with a label on it: Monday, Tuesday, Thursday, or Friday. Participants created eight sets of four baggies each. Each set, consisting of either all placebo or a mix of drug and placebo capsules, was placed in its own envelope, provided by the researchers and labeled with a QR code. Unknown to the participants, this code identified the envelope contents by the capsule contents and their order on each day. The study required only four of the eight envelopes, so

---

* Which sounds self-defeating and a lot to ask for from science.

† Regularly taking relatively small doses of a psychedelic substance, such as LSD. A small dose might be anywhere between a tenth and a twentieth of what people use recreationally.

the participants followed a sort of random process to choose the four they'd be using for the four weeks of the study.

The way the prep was set up, the partakers drew one of three possible combinations of envelopes: all placebo, all microdose, or half placebo and half microdose. Those who drew all placebo would be on placebo for the four weeks of the study. Those who drew all microdose would microdose for four straight weeks. The third group would have two placebo weeks and two microdose weeks.

At the beginning of each week, the partakers scanned the code on the envelope—communicating its contents and order of capsules by day to the researchers—and took their doses. They completed questionnaires about their experiences, and they guessed for each capsule whether it had contained placebo or psychedelic.

When the study ended and all was revealed, everyone had experienced improvements in psychological outcomes from baseline, including in anxiety, whether they'd ended up with all placebo or with a mix of psychedelic and placebo. Although there were tiny but significant effects distinguishing psychedelic over placebo, the study authors concluded that they occurred because the participants broke the blinding.

Indeed, 72 percent of the participants correctly guessed what was in the capsules they were swallowing. That's higher than the rate of 63 percent had they guessed at random, but lower than in studies of antidepressants (80 percent). The findings suggest that feelings of improved emotional state, anxiety, energy, and creativity* don't trace to the drugs themselves but rather to the *idea* that one might be taking the drug. In the words of the authors, the results indicate that the "anecdotal benefits of microdosing can be explained by the placebo effect."

I think there's another possible explanation for the results: the feeling of purpose and participation, of people paying attention to what you're doing, of being a part of something bigger and really interesting, even though it's illegal. Given that the placebo group improved despite not taking their usual daily microdose for four weeks, the psychedelic itself seemed to do not much at all.

---

* There were no improvements related to cognition.

A major caveat is that no one knows quite how big the dose was in the microdoses, as that was the user's choice. But because of the $N = 1$ design, each person was their own control, which helps mitigate that issue.

One of the participants who took placebo only had the best response: "An empty pill with strong belief/intentions makes nearly everything. You put spirituality into an empty pill here . . . wow!"

Other trials of drugs in this category, such as 3,4-methylenedioxymethamphetamine (MDMA), known to some of us as Ecstasy, have focused on people with PTSD. The trials often combine the drug with cognitivebehavioral therapy, not always with especially striking results.[21] One exception is the outcome of a placebo-controlled trial comparing MDMA plus standard therapy to standard therapy plus placebo. The forty-six participants in the MDMA group experienced greater improvement in PTSD symptoms than the forty-four in the placebo group. Even participants whose combination of PTSD and other conditions had proved resistant to treatment had better outcomes with MDMA. At the end of the eighteen-week study, 67 percent of the MDMA group no longer met the criteria for a PTSD diagnosis, which was significantly higher than the 32 percent in the placebo group who no longer did. Participants in the MDMA group also tended to reach this benchmark sooner.[22]

Some authors hypothesize that MDMA produces its effects by hushing an overbearing amygdala and turning up the volume on the more sensible frontal lobe.[23] Others have posited that other psychedelics work by quieting the DMN and boosting activity in the executive network.[24] But in at least one tiny trial, two of the five participants on placebo thought they were taking MDMA while undergoing more conventional psychotherapy. The authors concluded that the psychotherapy "could have produced a placebo effect" that led to the lack of differences between the groups. This suggests that it's not the chemical exerting the influence but the person's own brain.*

Placebo or not, the shushing of the DMN overall† could open the way to a softer border between self and other, you and the world, or even a

---

* As a neuroscientist friend of mine commented, "With a group of five, how would you even know?"

† Rather than the partial engagement of it while, say, reading fiction.

"decentering" of "you" that can drive anxiety.* These changes have been linked to . . . da da duhhhh! . . . changes in social cognition, particularly heightened empathy and caring less about whether or not you experience social rejection. Lower anxiety plus less attention to self equals increased empathy.[25]

With that equation comes the message that reducing anxiety isn't just about what you can do for yourself, but also about what you can be and do for and with others. A convincing placebo may yield the same results. Psychedelics do carry risks for bad outcomes in susceptible people, but I can at least comfortably suggest an empty pill capsule. And the data from these drugs are not that convincing. As one pair of authors put it, "Different results have been obtained with different methods,"[26] highlighting once again that unless scientists get their shit together and replicate studies more often, we'll never really know what's going on.

Trials of various over-the-counter supplements for anxiety tend to be tiny, to involve very specific groups of participants (those who are poststroke and who suffer from depression, for example), to be observational (able to show only a correlation), to have no controls, and to yield contradictory results or no effect.[27] A pattern I see again and again is a hint of promise in observational studies that evaporates in placebo-controlled trials.

BAKE AWAY THE ANXIETY?

Cannabis, once viewed as an audacious choice for self-medication, may be taking a back seat to psychedelics as the edgy intervention du jour to treat the burdens of our brains. As states have allowed for cannabis dispensaries to open and have countenanced the purchase of cannabis compounds in chocolate, gummies, oils, pill formulations, and baked goods, the allure of the illegal has dimmed.

I knew that things were taking a turn for cannabis when a relative in their seventies who has been a lifelong teetotaler suggested cannabidiol (CBD) to another relative for pain.† CBD is one of two compounds driving

---

* Although results are mixed, of course. They are always "mixed."

† Research seems to suggest that older people represent a large portion of the CBD-using population.

therapeutic interest in cannabis (the plant makes many compounds) and is thought to have no mind-blowing effects, unlike its partner THC, which is associated with such effects. It's possible that the two complement each other. But big, controlled, randomized trials are lacking, in part because of regulatory constraints limiting researchers' access to cannabis. The result is that the entire nation is one big laboratory, um, baking up results of personal trials of the stuff.

In addition to CBD's (perhaps unearned)[*] reputation for pain relief, among other purported benefits, it has a reputation for countering anxiety and stress.[28] Studies in animals suggest that it does do this. But the weird thing is that the effects tend to emerge only in animals that don't have ovaries (and the studies tend to be done only in animals without them). Furthermore, most CBD studies in humans have involved primarily people that don't have ovaries.[29] In rodents whose ovaries have been removed, exposure to estrogen or progesterone reduces anxiety-related behaviors. Rats, for example, show more avoidance of novel situations and higher levels of anxiety when their ovaries are removed. One manifestation of their avoidant and anxious behavior is burying marbles.[†] It's the rodent equivalent of suddenly becoming extremely focused on your smartphone screen to avoid making eye contact with—and then having to talk to—an approaching acquaintance.[‡] Rodents without ovaries turn to marble burying a lot more often than rodents with them, unless they are administered ovary-originating hormones. All of which suggests that having organs that make these hormones might set a high threshold for experiencing effects from CBD. For animals (including people) without ovaries, the threshold is lower and an effect may be detectable. Or, like interventions that show effects on global cognition, higher baseline need may predict a more obvious effect of the tool being used.

Randomized trials of CBD and anxiety are scant.[30] Some placebo-controlled studies have been conducted to test whether CBD has an effect

---

* I know that enthusiasts and true believers will yell at me for this statement, but the findings are mixed at best, and the studies are small and often lack controls. Pain has a huge subjective component, and I am a true believer in the important power of the placebo effect.

† Some researchers have questioned the accuracy of this interpretation.

‡ I am an expert at the human equivalent of burying marbles.

on anxiety, but they were small, one with five people per group and another with twelve per group (placebo versus CBD).[31] A no-drug-at-all study published in 2021 included forty-three adults who were asked to place an oil under their tongue that they were told contained CBD in one administration and didn't contain CBD in another administration. The oil contained no CBD in either case, but when the participants thought it did, they reported feeling more sedated, and they had changes in heart rate that were consistent with stress regulation.[32] A placebo-controlled trial involving thirty-two people who scored high for paranoia found no effect of CBD in decreasing anxiety; indeed, the trend was in the opposite direction.[33] Randomized trials are underway, too,[34] but perhaps dragging along. As an example of this drag, one of them was registered in 2015, had undergone seven protocol amendments as of 2019, was scheduled to terminate in June 2019, had a target enrollment of seventy-two participants, and still had not reported results as of spring 2021.

Another proposed trial, this one originally an FDA-approved, phase 2, open-label (meaning participants know what they are getting) study, started enrollment in 2015 to test CBD in people with anxiety, originally targeting an enrollment of sixteen participants to take sublingual CBD three times a day for four weeks. They were to be administered anxiety screens at the end of each week. Recruiting didn't begin until August 2018, and the protocol has been changed a few times, including removal of a plan to image participants and changes in the screening plan. In an April 2021 update to the study in a trials database, the open-label phase was reported to be complete and a double-blind phase to have been added. No results have been reported, so the world must wait.*

In 2020, a pair of authors evaluated the safety and efficacy of CBD in adults in a meta-analysis of twenty-five studies (with 927 participants total, which averages to only 37 people per study), twenty-two of them controlled trials and three of them observational (no controls).[35] Eleven of the studies assessed CBD and anxiety. Overall, the trials differed from each other in all the significant ways—for example, in terms of doses and dose regimens,

---

* You can read all about it here: https://clinicaltrials.gov/ct2/show/NCT02548559.

and types of formulation. The authors found some effects, more so in un-controlled trials, which signals that a factor beyond CBD may explain the outcomes.

In summary, of the eleven studies that evaluated CBD and anxiety (with only 358 participants total), two of them were the small placebo-controlled trials mentioned above, a couple of others found no effects of various doses of CBD on anxiety, one found a negative effect versus placebo (the volun-teers didn't have anything diagnosable, but they did test high in paranoid traits), and a couple found reduced anxiety but increased sedation. Two other trials that were not controlled found no effect of CBD versus placebo, and another found some effect. Another pair of randomized trials came to opposite conclusions: one "found no effect" and the other "found an ef-fect." Taken together, these results are just about as mixed as mixed can get. There's not even a discernible pattern.

The authors of the meta-analysis looking at these studies were unim-pressed overall, calling the results "promising" but deploring the "consid-erable variation" in how the studies were conducted. As long as researchers keep reinventing wheels with little tweaks each time and comparing apples and oranges, we will never be able to make decent comparisons or collect sufficient meta-analytical data on CBD's antianxiety effects, if any. Until then, the feeling of an effect, or its absence, is unverifiable beyond describing it as "only in our heads."

REAPPRAISALS

An important skill to learn to boost your stress cognition is cognitive reap-praisal. "Cognitive," of course, refers to "thinking" or "thinking through." "Reappraisal" means looking back at an incident and adjusting your per-spective on it. The idea is to reframe the stressful event in a way that leaves you feeling that you learned from the experience and were directed toward something purposeful. The parents of the children who died in the Sichuan Province earthquake made valiant efforts to do this.

Negative appraisals of stressful events—a totally typical reaction, don't get me wrong—have negative effects on us physiologically. There's nothing like coming away from a bad experience with zero traction on it, feeling as though nothing points us toward purpose. That total loss of agency leaves us disarranged, bodily speaking. The cyclic maintenance of our biological processes (homeostasis) gets disrupted as the stressor acts like a wrench in the machinery, impeding the works and throwing everything out of rhythm.[36] Our bodies love a good rhythm.

There are three general approaches to sharpening our stress cognition, and they interact with each other.[37] First, we can accept that there is no such thing as a stress-free life, and that stress is not a binary between "stress = bad" and "no stress = good."[*] As with any challenge we must meet, facing it with a few tools in hand can make dealing with it easier, which in turn can ease our reaction to the next stressor we encounter. We habituate to stress not merely by enduring it but by retooling maladaptive reactions to it because we have been there, done that.

Second, together with that acceptance, we can often evaluate the stressor as a challenge to grapple with instead of as a threat (unless it poses a genuine physical threat). This tactic relies on something like a mindfulness approach of objective assessment without judgment; we strip away the subjective coloring from our emotion, especially fear, as well as the biases we may be layering onto the experience. A cool appraisal can lead to a lower-temperature stress reaction, which in turn can result in less physiological haywiring. If you tend to play out scenarios, here's your chance to shine. You can imagine every possible scenario, good and bad, arising from a stressor—or an impending situation that's already causing you grief—and feel less of stress's first cousin, anxiety, about what's coming. Having the various possibilities laid out before you can be less anxiety inducing than staring into the unknown.

Third, once we've passed through the event—or, if it's chronic or ongoing, while we're dealing with it—we can practice reappraising it and its consequences. We can look for the "silver lining," even if the only shiny thread we can find is the fact that surviving a difficulty gives us skills and experience that might help us next time. This doesn't mean "looking for an excuse to explain away an outcome where I was at fault." You can't positively

---

[*] I mean, having no stress is a great goal for a getaway, if you can manage it. But it's not a tenable life-long lifestyle.

reappraise an incident as being "good" if your own negligence wrought it. But you can own your role in it, learn from the mistake, and move forward with purpose: a commitment that you won't do it again, that you'll make reparations, and that you'll absorb the lessons presented. These are positive responses, in the non-*kumbaya* sense.

If we take on stress with a mix of the three approaches, we can diminish the burden that some stressors impose. As we shrink the relevance and the perception of threat and find the sliver of value, we may limit the scars that stress can leave on body and brain.[38] We also can limit how much we allow stressing about our stressors to gouge those marks more deeply.

This process may sound like a "mind over matter" framing, one that opens the way to blaming a person overwhelmed by stress. It is not. Sometimes, all the skills and experience and reappraising in the world won't counterbalance the burden of our woes. We vary—both person to person and within ourselves from day to day—in how much balance we can maintain in the face of life's worries. That's why we also need other people to support us.

ANXIETY AND ACTIVITY

To hit (again) on a familiar theme, we've seen that physical activity is one of the few things most of us can do to free up cognition, social or otherwise. You're probably not terribly surprised to learn that it also can free up your reappraisal capacity because it dampens anxiety. An anxious mind is like a closed can of shaken soda—pressurized, fizzing, and chaotic. Motor effects are prominent in anxiety; think about the tweaky things you do when you're anxious, of how people pace or wring their hands. Movement is a relief, a way to pop the top on the can and relieve some of that unbearable pressure.* Afterward, you can better regulate yourself and appraise your situation more clearly.

If you can reduce anxiety by being physically active, then you create space in that defizzed brain of yours to use your stress cognition.[39] It's the

---

* In biological terms, exercise reduces levels of cortisol (the "stress hormone") and adrenaline.

circle of cognition: physical activity makes way for you to use your executive function efficiently for better cognition of all kinds. And reappraisal, as a cool-headed, mindful process, requires executive function. In a search for purpose, you have to attend to the stressor (attention cognition, the topic of the next chapter), deliberately dampen emotional input (social cognition), and engage in problem-solving (cognition cognition).

DE-STRESSING DIETS?

People have promised that low-carbohydrate/high-fat or high-protein diets* will do just about everything, but very rarely do such diets keep the promise. They don't "work"† for lots of reasons. When used for weight loss, the diets often simply end up yielding the same reductions in caloric intake as a lot of other diets because, as people who've tried them can attest, there's only so much fat and protein you want to choke down without carbs. Eating that way gets old fast, and any early weight loss that dieters sustain often reverses as adherence to the highly restrictive form of food intake wanes.

But. The one consistent benefit of such a diet, specifically the so-called keto (or ketogenic) diet, has been to reduce seizures in some people with epilepsy.‡ And there are hints that a low-carb, high-fat diet could take the edge off anxiety. Hints.

What purportedly happens in the brain is related to levels of glutamate—the neuron-*exciting* neurotransmitter—which is depleted in the keto diet.[40] Glutamate plays a vital role in anxiety. The yin to the yang of glutamate, if you recall, is GABA,§ the neuron-*inhibiting* neurotransmitter. People who have diagnosed anxiety and panic disorders have fewer places for GABA to bind to their neurons in some parts of their brains That means fewer places where GABA can counteract excitation.

---

* The commonality is the "low-carb" part.

† I am not a fan of "diets" undertaken for nonmedical reasons. I like to eat what makes me feel pretty good in body and brain and to not eat what doesn't.

‡ For more about the keto diet for epilepsy, go here: www.epilepsy.com/learn/treating -seizures-and-epilepsy/dietary-therapies/ketogenic-diet.

§ A quick mnemonic refresher: GLUE = GLUtamate Excites; GABI = GABA Inhibits.

Certain drugs used to treat anxiety and depression are in a class called selective serotonin reuptake inhibitors, or SSRIs (you probably recognize the most famous of these: fluoxetine, or Prozac). As their name suggests, they keep cells from soaking up the neurotransmitter serotonin once it's released into the space between neurons. The result is that the serotonin persists, so its message persists, and its message is allegedly to "smooth your mood."* In addition to this effect, SSRIs boost GABA levels in the brain,[41] suggesting that GABA might have a role in taking the edge off a negative mood. Those Xanaxes and Valiums that people talk about all the time are part of a drug class, the benzodiazepines, that operate on the GABA-signaling system.[42]

What's all this got to do with eating low carb/high fat? A ketogenic diet boosts GABA and dampens GLUT, which suggests that it exerts some reducing effect on anxiety and perhaps a leveling of mood.[43]

Some authors have argued that the keto diet "remodels" the microbial populations living in the gut, affecting what gets to your brain and what your brain uses as fuel.[44] The microscopic organisms in your digestive tract break down what you send them into new products, or metabolites, some of which go to the liver for processing. The products from the liver determine the profile of molecules that reach your brain, where they are used for building, communicating, and fueling the brain's functions. Astrocytes, which are the purveyors of energy molecules to neurons, play a role in how metabolites from the keto diet are used.[45]

But no one knows for sure that it all goes down this way between the gut microbiota and the brain—it's early days yet in tracing these pathways. And it's hard to avoid eating carbs, which makes sense because many of your body's systems rely preferentially on carbs to run. Furthermore, some kinds of fats in particular are risk factors for cardiovascular disease. Finally, it's hard to stick to a healthy keto diet. People with epilepsy who follow the diet therapeutically have specialist dieticians map out an eating plan for them.

---

* This proposed explanation for the mood-stabilizing action of these drugs has met some obstacles that have deflected it toward other, more likely candidate pathways, some involving remodeling connections between neurons.

MIND YOUR MULTITUDES

Your brain contains multitudes in the Walt Whitman sense,* different facets of you that are always present, the parts that make up the whole you. What are their roles? If you look without judgment and use perspective-taking on your multitudes, maybe you can understand why they operate the way they do. This concept arises from an approach called internal family systems therapy,[46] developed by Richard C. Schwartz, that has worked for some people to relieve burdens of anxiety and stress. But like all such efforts, it doesn't work for everyone equally or sometimes even at all.

The process involves recognizing your multitudes, the residents in your head that you view as troublesome, and beholding them without judgment and with open compassion. It's like empathy for the parts of you that bother you most. If Anxiety is present, look at it. Why is it there? What is it trying to do to be useful for you? And without crushing Anxiety's feeling of usefulness, see if you can put it into a "waiting room" within your mind to take a breather while you function.

It's possible that as you refine how you approach your Anxiety and its reason for occupying space in your brain, you can use some of the other techniques covered in this book to send away it and its companion, stress: physical activity, reappraisal, and playback of storytelling (see below).

If this seems like an invitation to entertain the possibility that you have several personalities, it's not. The idea behind internal family systems therapy is to recognize that your psyche comprises different facets that serve different purposes, and to study those facets compassionately. After all, if you want to practice compassion with others, you might be well off practicing on yourself first.

Your Anxiety isn't there to hurt you. It's there to protect you. It has been baked into the human brain over millennia of evolution. The high arousal and danger calculations that it motivates keep you alive. It is almost parental in its urgency to protect you. Beholding your Anxiety as a parental voice motivated by the need to keep you alive—instead of as a monkey tightly gripping your back—might make it easier for you to move Anxiety into the

---

* From *Leaves of Grass*, "Song of Myself," section 51: "Do I contradict myself? / Very well then I contradict myself, / (I am large, I contain multitudes.)"

waiting room of your mind palace. Gently saying, "I must go now" to that part of you is probably easier than peeling an angry, grasping, freaking-out primate from your back.

You can also try this strategy with thoughts that you'd rather not deal with, such as catastrophic thinking or intrusive ruminations. Look at the thought not through moral valence or emotion, but from the perspective of how it compares to other thoughts. Like your multitudes, a thought that you don't want running the show can be placed in context, viewed as one of a set of possibilities rather than the dominant one. Sure, catastrophe is a possible outcome of many scenarios, but it's not the only one. You can summon alternatives if you practice your prediction skills.

Whether the approach of internal family systems therapy resonates with you or not, the mindfulness aspect remains useful against stress and anxiety.[47] Some researchers claim that therapies incorporating mindfulness work just as well for anxiety as traditional cognitive-behavioral approaches do. The latter emphasize change, whereas mindfulness focuses on acceptance, but you can do both. You can reappraise events and, if needed, change future behaviors with a sense of purpose—all while remaining nonjudgmental and compassionate with yourself.

You will probably not be surprised to learn that mindfulness also shows benefits against stress. If there's any group on Earth that endures prolonged stress as an everyday reality, it's medical students: they're ambitious and overworked, and they regularly take people's lives into their hands. The year-plus coronavirus pandemic has translated into even more stress than usual for medical trainees. For this reason, they are frequent targets of interventions for stress, including mindfulness. A 2021 study of 143 of these beleaguered (usually) young people found that those who engaged in eight two-hour mindfulness sessions during the sixteen-week study period showed "clear improvement" in their self-perceptions of stress and symptoms of anxiety and other conditions.[48] The sessions did not, however, do anything for their feelings of burnout, which is a systemic issue that remains to be addressed. More broadly, mindfulness-based interventions seem to have the greatest effect on stress.

There are many different ways to access a mindfulness practice, with enhancements from nature or psychedelics or music you like,[49] and they are highly individual. I can't recommend specific sources for you because what works can be so personal. But there are apps for solo-guided techniques to practice on your own, and you can check local practitioners if a class or guided sessions are more your speed.

REAPPRAISAL THROUGH MOTION AND STORY PLAYBACK

I interviewed a young woman whom I'll call Sarah about her experience with another therapy that targets stress reduction, especially as it relates to trauma: eye movement desensitization and reprocessing, or EMDR. She told me that after four months of talk therapy to identify what to work on, she underwent her first round of EMDR, with striking results.

To prepare for the EMDR sessions, the therapist asked Sarah to choose a safe place in her mind where she could go if things started to feel uncomfortable. She had to practice going there a few times a day so that it became easy for her to do. During the sessions, Sarah held a pair of sensors that vibrated off and on as she related a deeply distressing event. The therapist in turn related the story back to Sarah while the vibrations continued at Sarah's fingertips.

"It's very emotional, and I didn't expect that," Sarah told me. "It made me feel really wired, it made me tear up, I was getting shaky, and my brain was straining." She described it as feeling a sort of "pressure from the inside."

The effect on her appraisal of the stressful event was immediate. "I remember opening my eyes, and it was like, I don't know, I was, like, happy," she said. "I took a deep breath and I was like, 'Wow.'"

After three sessions of EMDR, Sarah said she felt finished because it did what she needed it to do. About the triggering event, she said, "It feels like an event that happened, no emotions tied to it. It used to make me really upset."

EMDR was developed in the 1980s for PTSD. The idea behind it is that performing specific eye movements during the telling of a traumatic memory somehow helps the person store the memory differently. During memory re-creation through storytelling and body movement, a trained therapist moves their fingers in front of the client's eyes. The client is instructed to follow the fingers with their gaze, or, in some scenarios, to pay attention to rhythmic tones or vibrations made by the therapist.

The stimulus is "bilateral" in that it involves sensory input on both sides of the body, often alternating. The experience of this stimulation during the retelling of a traumatic event hypothetically dampens the emotion and the clarity associated with the memory. Later in the session, the therapist leads the client in a guided "walk" away from the memory and toward more pleasant recollections.

It sounds fabricated. Why would shifting your eyes around while you recount trauma have any effect on your emotions about or memory of that trauma? Yet EMDR has been viewed as effective enough to be used for anxiety about dental procedures, and in one small study it was found to be equal to cognitive-behavioral therapy in its benefits for people with COVID-19-related trauma.[50] No one has any idea how it works, and many experts are not convinced that it does. The person who developed the technique, Francine Shapiro, came up with it after realizing that she felt better after a nature walk during which she frequently moved her eyes to take in the beauty around her. You'll notice that she was both out in nature and moving, which could explain the easing of her negative feelings. But Sarah was in an office, going through the procedure in a clinical setting.

Some contrarians have suggested that what actually works about the practice are the elements it has in common with cognitive-behavioral therapy—specifically, recall and the entertaining of more positive thoughts. But studies suggest that the movement and rhythm aspects of EMDR matter somehow.[51] The therapist's tactic of relating the story back to the client could be a factor too.

In a meta-analysis published in 2020, the authors wrote that EMDR "may be effective" for PTSD, but that the quality of available studies was too poor for them to stake a reputation on that conclusion.[52] Nevertheless, when they compared the effects of EMDR to those of cognitive-behavioral therapies, they found no difference for anxiety—suggesting that EMDR, however it works, performs as well as standard approaches. Some professional societies, including the American Psychological Association,* accept the effectiveness of EMDR enough to incorporate mentions of it into their guidelines.[53]

SEARCH YOUR FEELINGS

The ability to recognize what you are feeling can be important to the self-evaluations we've talked about in this chapter. It's hard to behold an emotion or put it in a waiting room or make it just another part of a collection of feelings if we can't recognize that we're having the emotion in the first place. Judith Moskowitz told me that part of coping is this ability to recognize feelings. In her work, she has found that when people simply logged their daily emotions, they experienced more positive emotion. "So there is something about increased awareness of emotions and being able to actually label them," she said.

She mentioned a study from the 1990s by George Bonanno, professor of clinical psychology at Teachers College, Columbia University, and an expert on coping. Researchers videotaped interviews with bereaved spouses as they spoke about their deceased loved ones. They then showed the videos, without sound, to observers who had no idea what the interviewees were talking about. The naïve viewers were asked to rate each interviewee on how likely they would be to ask for help from them. They gave higher ratings to people who smiled while talking about their dead spouses.[54] "Even though they were talking about something painful and difficult, they showed some sort of positive emotion, and brought in this social support," Moskowitz said. "It's a nice demonstration that in the midst of expressing something really difficult, if you're able to show some hints of positive emotion, you're more likely to draw people in."

Had the naïve participants been able to hear the stories, they might have had a different reaction. We've already encountered the potential benefits of storytelling for building cognitive empathy. But it has stress- and anxiety-relieving benefits too.

Hurricane Harvey struck Houston, Texas, in 2017. More than a hundred people died, and the storm left behind a trail of damage and loss that

---

* Which "conditionally" recommends EMDR for treating PTSD.

lingered for some individuals in the form of PTSD. A group of researchers wanted to know if personal storytelling might benefit these survivors, and they chose an unusual way of testing the question. An improv activity called Playback Theatre, introduced in New York in the 1970s, relies on actors to hear personal stories from the audience and then act them out onstage. Playback companies have been established around the globe.

A performance involves a master of ceremonies, a musician, a handful of actors, and, of course, the audience. The MC asks audience members to share something true about an experience from their lives, which the actors, accompanied by the musician, then reenact. As with all improv, the brief is to always move forward with a positive. The power lies in the authenticity of the stories and in honoring the original speaker during the playback. A gift is provided in this in-the-moment resonance around a stranger's truth and in the storytelling. The call-and-response aspect of the event is also a feature of the EMDR experience that Sarah described, relating her story and then having the therapist retell it to her.

Based on reports that Playback Theatre seemed to offer benefit for people with trauma, researchers assessed whether a similar process might reduce anxiety and PTSD symptoms for thirteen survivors of Hurricane Harvey. They also conducted brain imaging on the participants before and after the intervention.[55] The participants reported decreased anxiety and PTSD symptoms, but no effect on their depression. Given the autobiographical nature of the experience, as the original narrator observes their own story being played out, you can probably predict what the MRIs showed. They showed increased amygdala connectivity to a part of the parietal lobe involved in language and emotional response. This region in turn was linked more strongly to the salience network and the DMN after the intervention.

The actors may also have experienced a decline in stress or anxiety. Research suggests that behaving prosocially, in a way that helps others, offers benefits for well-being, anxiety reduction, and relief from mental distress. To quote one study, "Doing good feels good."[56] Even better, perhaps, "doing good for others feels good."

Although "self-care" is important, evidence suggests that behaviors focused solely on ourselves can actually lead to increased anxiety. Perhaps the pressure to find time to stare at a candle or gaze into the middle distance while emptying the mind instead turns up the volume on the DMN. This association raises the intriguing possibility that self-care could sometimes mean engaging in caring activities for other human beings.[57] The trick is to find the balance: avoid overdoing either self-focused or other-focused activities.

Speaking of balance, I think it's important here to bring up the idea of resilience. A notion that has gotten a lot of attention, resilience is framed in some cases as a personality trait that either we have or we lack—but that we all need. If we don't have it, we may be sold various ways to try to build it.[58] The problem with framing resilience as a trait (born with it, immutable, fixed), rather than a state (situational, changeable), is that people who can't find resilience in a given situation might be blamed for not having it. Resilience is situational. Environment determines how much of it we can access and when. Support, including social support, feeds and sustains it. As one researcher put it, "social and physical ecologies" determine the depth of our wells of resilience. Resilience is a state, not a trait, and none of us can depend on having unlimited reserves of it at all times.

In a Zoom seminar on mental health and the pandemic, sponsored by *Undark* magazine, Bonanno, the expert on coping mentioned above, listed another dimension of resilience: time. "Something happens, and we're resilient to it or not," he said. "That's a matter of time, an outcome, the result of what we've done." Some people take longer than others after a traumatic or unsettling event to regain the contours of self and their feeling of purpose, and some people never regain it. Furthermore, Bonanno added, "Being resilient to one event doesn't guarantee resilience to the next major challenge." Nothing ensures resilience, he said. Of all the candidate factors that might contribute, including a limited role for personality, Bonanno pointed emphatically to "social resources."

Judith Moskowitz, who also participated in the seminar, talked to me a little more about resilience. "The problem with saying 'that person is born resilient' is that it implies that they can't change, that people who aren't born resilient can't become resilient. That's why I bristle at that characterization."

So, then, what is resilience, and where can we get some? Moskowitz said, "I think of resilience as the ability to make it through difficulties and sort of bounce back. Get through it, get back to being OK. I would call that being resilient." And, she added, you can learn it. You can practice coping skills and gain the ability to practice resilience.

Resilience doesn't mean flexing back to normal. It's recognizing reality for what it is, without judgment. It means acknowledging what is before you, and taking it as it comes and goes.

CHECKLISTS

Three-Step Stress Management

1. Accept that a stress-free life does not exist, and stress is not a binary of "stress = bad" and "no stress = good."
2. As much as possible, evaluate stress as a change or a challenge to address, rather than a threat.
3. Practice reappraisal.

Additional Strategies

✓ Breathing exercises
✓ Reappraisal
✓ Doing something good for another person
✓ Finding purpose, big and small
✓ Physical exercise
✓ Being in nature, big or small
✓ Mindfulness
✓ Having empathy for your multitude of "parts"
✓ Sharing stories

CHAPTER 7

ATTENTION AND MEMORY COGNITION: GAINING FOCUS

I F YOU DON'T remember your dreams often, that may be because you don't wake up enough during the night. You might think that means you don't wake up enough during rapid eye movement or REM sleep. But according to dream specialist Perrine Ruby, a researcher at the Lyon Neuroscience Research Center, we dream during all parts of our sleep cycle. True, we are most likely to have a "dream report," a recollection of something we've dreamed, if we waken during the REM stage. Some people even have dream reports if they waken upon entering sleep, well before they move into their first REM stage of the night. The waking is what may allow you to create a memory of a dream that you then store to become a lasting memory.

Many of us have had the experience of trying to capture a dream as it melts away. That's the "whiteboard of the brain," our working memory, losing the trace of the dream events. Many of us also have had the experience of grasping at some tiny remnant of the evanescing experience that, if we can hold on, will pull us back into remembering the whole story. If we succeed in that recall, we create an episodic memory. We pack away features of the dream that, if we wish, we can later reassemble as a memory. The duration of these lasting memories can be powerful. One of the earliest dreams I can recall is from the early 1970s. I still remember (or at least reconstruct as a memory) some parts of it quite clearly.*

In a small study, Ruby and her colleagues found that people who are "high dream recallers" had relatively high DMN connectivity right before waking up. That might have helped them keep the Magic Marker on the working-memory whiteboard a little longer.[1] But on average, we experience a dream recall about once a week, Ruby told an audience in a 2016 TEDx Talk. In our dreams, she said, we enter a virtual world where "anything is possible." And to illustrate that "anything," she showed images drawn by people recalling their own dreams. Among them was a man who'd removed his own head (in his dream) so that he could see behind himself. Clever.

We can also discover a solution to a problem or untie its knots during dreams. And a few of us, maybe one in thirteen, can experience what is known as lucid dreams, in which we are still asleep but aware that we are dreaming and can take control over what's happening. For Ruby, the "democratic miracle" of dreams is that just about any human brain can access them; as such, they are places that are "free of the constraints of the waking world."

But the waking world can certainly invade our dreams. In fact, as we contemplate and evaluate waking events in our dreams—whether we remember doing so or not—we are engaging in a form of memory consolidation. In this process, we coalesce features that (presumably) are most salient for us to box away in case we need to reassemble them.

In 2019, Ruby and her colleagues did a study of thirty-two high dream recallers to evaluate the potential link between what we dream about and how it might support creation of memory.[2] The participants underwent exposure to "episodes": various combinations of odors, visualization of locations, and imagery. Then they slept, were briefly wakened at five a.m., and then were allowed to sleep some more, until their regular waking time. After that came the dream reports.

---

* It involved a cartoonish, oversized whale hanging out in the driveway of my grandmother's house in Waco, Texas.

A whole lot of them dreamed about the experiment itself and their participation in it. "I dreamt that I woke up to report my dream using the voice recorder just as in the experiment," one said. Another described "guys in white coats" who were "doing experiments on us." Not all dreams with lab-related flourishes were quite as predictable. One participant reported, "I was in the building for the study of dreams, downstairs in the cafeteria, and I was explaining to a rhinoceros that I was preparing my dreams as my handbags, with many objects which could be useful." Those who had lab-related recollections or who recalled dreams about something they'd learned also had better visuospatial recall of the episodes they had explored while awake. This better recall suggests that their dreaming helped them solidify the waking experiences.

What would be fantastic is if we could leverage this consolidation capacity and the lucid-dreaming ability some of us have to improve our learning and memory. Unfortunately, the methods (drugs, cognitive techniques, various stimuli) that have been tried to achieve this, um, dream have yet to work reliably or consistently.[3]

If you can recall your dreams, you probably can do so with a strange level of detail, like the high recaller who explained their handbag-based plan to a rhinoceros interlocutor. The organ that makes this possible is one we encountered in Chapter 1 as we peeked underneath the temporal lobes: a pair of saddle-shaped structures,* called the hippocampi, and their close companions, the paired amygdalae. We know that the hippocampus is important in dreams in part thanks to studies in people whose paired structures have been compromised in some way.

In one such study, the four participants had amnesia along with their damaged hippocampi.[4] As a result, they had fewer dreams. The dreams they could recall lacked specifics, including spatial and sensory details. Sample recollections were along the lines of "just walking up a railway line." One respondent reported, "I was in a nightclub. I was drinking. People were dancing," with no addition of detail, even after prompting from the interviewer.

---

* If you look at the hippocampus and its neighbor, the fornix, together, you might be able to visualize a seahorse, with hippocampus as the tail region. This appearance is how the hippo (horse) + campus (sea) got its name.

Meanwhile, the ten participants with intact hippocampi reported dream memories that were rich with detail. One recalled dreaming of an effort to improve the Sunday edition of a well-known newspaper that was being sold to another company. The dream took place in a large "restaurant-type pub-type building" that was "packed," and there was some kind of bidding war for the paper. The people present were looking at individual screens, possibly with bidding amounts on them, and an auction was about to begin.

That's a lot more detail than "I was in a bar. I was drinking," and the hippocampus may be why. The difficulties that those with damaged hippocampi had in forming and recalling memories in waking life appeared to influence their ability to dream and process details. Dreaming and memory seem to share a reliance on these two seahorse-tail-shaped structures deep in the brain. They probably also share a reliance on attention, which guides the brain like a beacon to illuminate what's important (or at least, in dreams, what seems important, like a chatting rhinoceros), which is the beginning of creating memory.

As you'll see in this chapter, paying attention to our attention and being smart about understanding how we use it can be one of our best memory aids. Attention cognition is the gateway to all of our other cognitions.

MEMORY IN A BOTTLE

In the mid-1900s, neuroscience went on quite the roller coaster ride in its quest to discover more about memory. First came the proposition that some single molecule might be responsible for encoding memories, or "engrams." Imagine if that were true. Memory in a bottle, reconstituted synthetically from this mystery molecule. One candidate was RNA. Given the way neuroscience operates and what we're learning about the increasing complexity of this molecule and its cellular roles, it may one day step into the limelight, directed once again toward memory research.

But in 1966, researchers from several different labs had experimented with the idea that RNA might play a role in learning transfer, and had met with no success. Not ready to let go of the "engram is encoded by a single molecule" idea, they floated another proposal. This one posited that short strings of amino acids (the building blocks of proteins), rather than short strings of RNA nucleotides, might be the key molecules in embedding memory.[5] Unfortunately, after several years of hand-waving about one candidate in particular, called scotophobin because it seemed to make rodents avoid the dark, that hypothesis got torpedoed too.

As you can see, neuroscience is wreathed with unexpected twists and turns. Something can be considered "fact" for years, even decades, until suddenly it is fact no longer. Kind of like our memories as we age—or as we come out of a delicious, fleeting dream.

The engram is having a revival of sorts thanks to studies that have expanded from looking for a single molecule to looking for ensembles of neurons responsible for re-creating events in our lives (or our dreams). Rodents have proved quite valuable,[*] allowing us to investigate which collections of neurons are involved in forming and accessing their memories related to, say, fear, as an example.[†] Turns out, the way rats "freeze" when they are afraid is so clearly a fear response that we can use it to trace everything underlying it, down to the level of the cells involved.[6] This information is useful for humans because it's a baby step toward figuring out the pathways in people who involuntarily replay fear-related events in their heads, as may happen in PTSD.

We've learned that the amygdala contributes its part to memory. It works in concert with the hippocampus to make memory more than just an instant replay pulled together through chains of neurons. It helps reconstruct memory painted with emotion. In rats, the emotion we work with is usually fear, because we have limited insight into their inner lives. But for us, emotions are a range, a spectrum, wide and deep, with many layers and co-occurring states.[7]

For long-term memories to form and be retrieved, the events involved must first be limned on our working memory's whiteboard, supported by our prefrontal cortex/frontoparietal system, including executive function. This setup tells you that the attention network—choosing and ordering

---

* To which I say, yet again, "Thanks, rats." I like rats, and the impending possibility that I might have needed to use them for research is one reason I left the bench.

† It is worth noting that the sets of neurons that combine to replay memories don't always consist of the exact same neurons. There is some flexibility or redundancy there.

what's worth putting on that board and what isn't—is important to memory. If working memory is a whiteboard, this system is the self-guided marker we use to record on it.[8]

A pair of subsets of the frontoparietal system may work as alternate toggles to regulate communication between attention and the DMN. They can flip the connection off or on, potentially determining how much capacity our working memory has.[9] This capacity—the size of the whiteboard—is associated with measures of intelligence, implicating the DMN and its role in these shifting communications, and therefore in global cognition.[10] Like our measures of global cognition and like our DMN connectivity, working-memory capacity changes over time, increasing through childhood and plateauing in late adolescence or early adulthood.[11] The catch is that although we can expand our working memory through various techniques and in really specific ways, this capability doesn't translate to generalized improvements in cognition.

If we are to retain what is recorded before it disappears, we have to successfully transfer and store the elements of the memory where we can retrieve them later. The whiteboard space is finite, and if we lack capacity there, we won't pick up new information and store it either. No room. Our hippocampus, long known for its role in episodic or long-term memory,[12] contributes to this process. It syncs up with the oscillations in the cortex, where we pack away information we are trying to remember.[13] The connections between the hippocampus and the cortex become stronger as we add to the working-memory load, with the hippocampus valiantly maintaining the ink on the whiteboard a little bit longer. Studies indicate that the hippocampus and cortex communicate in one direction when the information is being input and in the opposite direction when it is being recalled.[14]

If our working memory is oversubscribed, we (and our cortex and our hippocampi) are oversubscribed.[15] You have no doubt experienced this sensation, perhaps when many people are speaking at once while you're trying to read or to tackle a complicated problem. The brain has a finite amount of energy available at any given time. When your brain feels like it's full of buzzing bees, you're running out of those resources.

If we can get to the point of storing a memory, our capacity to do so is extraordinary.[16] It's what we access when we want to generalize lessons learned from early experience, engage in familiar activities, and mentally time travel into the past. In this way, the memories that we store are key to the process of learning.[17] Some stored memories relate to personal events, or episodes, and are episodic. Others are just bits of information (called semantic memories), whereas still others are memories of how to do things and are more procedural.[18]

Forming these procedural memories creates an autopilot that we can initiate without even thinking. We can turn on the stove and place a pan on it in readiness to cook without the need to pause at every step and consider what we're doing. The DMN may play a role here, delivering these unconscious standard operating procedures for environments that offer no new information, freeing us up to think about other things while we put water on to boil.[19]

When we unconsciously activate this procedural memory from the cortex, we use "top-down" processing, defaulting to everyday habitual activities.[20] These general information frameworks are what let us pull on a T-shirt almost unconsciously. But say one morning, we find a T-shirt with the bottom sewn up. Now we have something unfamiliar to deal with. Instead of following our unthinking "throw on a T-shirt" procedure while we contemplate coffee, we have to focus on the shirt. We have to consider how to get the shirt open while we wonder, "Who the hell stitched up this shirt?" Doing so engages bottom-up processes: we see something wholly new (unless unexplained wardrobe defacement happens to you a lot), and we must attend to a number of cues to figure out what's happened. We've switched from autopilot to manual transmission while we work on it.

So when we're cooking a dinner we've made many times before, one part of our attention network, the dorsal attentional network, alerts us only to important items, such as a spice we need to find. Meanwhile, we default to autopilot for the parts of the procedure we've performed frequently. But if a grease fire starts while we're looking for the oregano, yet another attentional network, the ventral frontoparietal network, will tap the dorsal attention network on the shoulder and say, hey, forget the oregano, the stove is on fire.[21]

THE SOCIAL MEMORY

As the engram experiences its engrammatic comeback, neuroscientists are bringing up the concept of a "social engram," with the hippocampus working in the development of social memories. Our hippocampus may be the little social engineer in our heads that allows us to see and remember one another as human and treat each other that way.[22]

It also may play into our tendency to engage in neural replay, a process we execute constantly, during sleep, waking, hanging out, lying still, and resting.[23] During these replays, spontaneous spiky waves of synchronized oscillations wash across our brains, reviving patterns of activity for specific details from our memories. These back-and-forth waves involve the DMN, which toggles off and on as they occur. When we're awake, this activity produces what we think of as introspection.

What we are doing with these replays is inspecting their elements, which you've likely experienced as checking out of your surroundings and turning inward. In the middle of a Zoom meeting, you may have found yourself reviewing your role in some unrelated event while a colleague runs through a deck of dull PowerPoint slides. You conduct the tasks in parallel, looking at the slides, sure, but also replaying the memories, often social ones. And then you click out of the replay when something in front of you pulls you back to the present. Did you miss anything important? Evidence from a 2020 study suggests that while your DMN was busy serving up those memories, alpha waves were busy inhibiting your attention to other things,[24] so . . . possibly, you did. But at least you know why, neurologically speaking.

Perhaps you caught the tail end of a part of the presentation that clicked on your salience and executive function networks. If so, you may have captured enough on your working-memory whiteboard to hold it there, allowing you to backtrack so you look like you were paying attention the entire time. This process feels a lot like capturing that vanishing thread from a dissolving dream. Sometimes you catch just enough to pull yourself back in and remember the rest. Sometimes it is too elusive and you lose it.

What you tried to do is shift the dominant cognitive system—in this case, telling your DMN to hush so you could toggle up your attention and working memory. If it didn't work and you failed to capture what was going on, you might even apologize: "Sorry, I wasn't paying attention for a minute there."

Attention is the first step in memory. We can't write something on our mental whiteboard if we don't notice it in the first place. In an optimal existence, attention operates to keep us safe. It orients us to important information in our environment and, at least in theory, keeps us from wasting our precious reserves on useless information and overloading our whiteboard. As you read this paragraph, your environment is packed with countless sensory cues—visual, auditory, tactile—inside your body and out. How many of them did your attention illuminate while you were reading? Probably very few, because they aren't relevant to the task at hand.

If attention's reconnaissance pulls us off mission and away from important information, then our attention may be operating poorly. That's not to say that attending to apparently inconsequential cues from the environment is a waste of time. We may notice things that others don't, and that can be serendipitous. From there we find new discoveries and insights or even a new understanding of something familiar in an unfamiliar context.

Ask your friends and family what they think about their capacity for attention, and you'll find that many people are dissatisfied with their skills in that arena. They feel that they can't focus—can't shut out environmental cues. Or that they are prone to checking out of the bigger picture by homing in on one thing. Or they are prone to delving too deeply, being overwhelmed by their DMN and internal personal audits.

The question is, can they do anything about it? The answer, you'll see below, is yes. You can employ strategies to take the driver's seat of your attention. You can leverage your working-memory capacity to its fullest extent. As for episodic memory, that's a different story, and one that you likely can reinforce only by continuing to tell your stories.

We saw in the last chapter what stress can do to our thought patterns. We will see in the next what mood, which we owe in large part to our DMNs, can do to it as well. When these systems degrade the power of our attention, we lose power in our working memory too. We make mistakes, we feel worse, we turn inward, our mood darkens, and we can create a cycle. But you can break the cycle.

An important step is awareness that the cycle exists and knowing what its components are. As you learned in the last chapter, for example, reevaluating a stressor can be useful in reframing how it affects you, mitigating its ongoing damage. As we also saw, you can discharge some of the cognitive load of stress to create mental space and make your attention less chaotic—to use your stress cognition together with the attention cognition approaches you find here.

THE SUPERATHLETES

Every year, the best athletes in their sport gather for a championship match in one of the capitals of the world (although in 2020, the twenty-ninth championship was virtual). In brutal and intense bouts, three hundred competitors from more than a dozen countries go head-to-head—or, more specifically, brain-to-brain. They are the memory athletes, and how they perform incredible feats of memory is both surprisingly straightforward and shocking. The winner in 2020 was Emma Alam, a young woman from Pakistan, who shattered a few world records while she was at it.

Dominic O'Brien, of Great Britain, has won the championship eight times.[25] As a child, he sustained a "knock to the head," had dyslexia, and experienced what he believes were probably attention deficits, accompanied by great difficulty memorizing things. His school reports didn't augur a career as a memory athlete. His teachers described him as "terribly slow" and "(mentally) absent"; perhaps most ironically they wrote of him that "he cannot keep a question in his head."

O'Brien was mired in the business of extracting silver from X-ray films, which is about as messy and unexciting a job as it sounds, when he saw a British mnemonist (someone who memorizes long lists or strings of things) on a television show in 1987. O'Brien, age thirty at the time, watched as the mnemonist, Creighton Carvello, demonstrated his ability to view an entire deck of cards, one card at a time, and then recite what he had just seen, quickly and perfectly in order. Hooked, O'Brien became good enough at the trick to be invited to his first memory championship.

At these competitions, the athletes have to perform such tasks as memorizing a two-thousand-digit number. They get an hour to review the digits and an hour and a half to create the memory. Or they might have fifteen minutes to memorize 180 unfamiliar faces in order. Or to memorize a list of three hundred words in order.

Are they neurologically special? A group of authors assessed the brains of some of these athletes, including O'Brien. They found that structural differences didn't explain their superior skills, and neither did any especially robust intelligence capacities. What stood out was how they used specific regions of their brains that are important to memory. Spatial memory, which the hippocampus strongly supports, was a key.

OK, sure, whatever, you're probably thinking. These people have time or motivation or interest or all three, and you don't.

Consider the results of a 2017 study using MRI to image the brains of twenty-three memory athletes, matched for sex, age, and intelligence test scores with nonathletes who undertook a six-week intensive memory training, and active and passive controls. The researchers found that the training shifted the brains of the nonathletes to a pattern of activity similar to that of the athletes' brains.[26] Furthermore, the memory improvements lasted up to four months after the training ended. Imagine becoming sort of athlete-like after a mere six weeks of training. That's not something you can do by, say, training your quads in the hopes of becoming a competitive sprinter.

Tests of the memories of regular folks are a little less rigorous. You may recall, depending on your own memory capacity for the events of 2020, the episode of Donald Trump's memory test. In an interview on Fox News, Trump reported that he'd "aced" a memory exam that required him to remember a series of five words to repeat when asked to do so a few minutes later. The five words Trump cited—"Person, woman, man, camera, TV"— quickly became an internet meme. Trump claimed that the clinicians administering the test were so astonished by his success at recalling the words that they gushed, "That's amazing, how did you do that?"—which seems rather doubtful.

The test Trump was apparently describing is the Montreal Cognitive Assessment, or MoCA. A screen for dementia used in a clinical environment, it consists of thirty questions. The delayed recall portion assesses whether someone can keep five words on their working-memory whiteboard and recall them after a few minutes. The test taker first repeats the five words immediately after hearing them. The recall request is made after some other parts of the test are completed.

The request to recall the words is not an ambush: the test administrator lets the test taker know that they'll be asked to recite the words at the end of the screening. The words aren't usually associated in the way that "person, woman, man" and "camera, TV" are, so Trump probably took those examples from his immediate television interview surroundings. The five words would more likely be unassociated, such as "nose," "cotton," "steeple," "sunflower," and "purple."

When people test working memory in a research environment, a commonly used instrument is the N-Back test, a version of which gained some fame for potentially offering global cognitive effects. If it does have any such influence, it's small, at best.[27] Results, as they say, are inconsistent, which is another way to say that researchers all conduct their studies differently and no one's getting the same results.[28] The task takes its name from the use of "N" in science to indicate "number of things," as in "we included all people with brown eyes (N = 6.5 billion)." For N-Back, N = the number of items presented in a series before the item repeats. If N = 1, then the item repeats immediately after it appears, as in B . . . B. If N = 2, then it repeats two steps away from its initial appearance, as in B . . . T . . . B. The higher the N, the longer the string of elements (or images or shapes or whatever's being used), and the faster they are presented—making the test harder as it goes.

The test involves loading up that whiteboard of yours, even as you struggle to keep the ink from fading so you can check the board and recognize a repetition when it happens.[29] But you're also erasing items that you don't need anymore: once you've pegged a repeat, you can dispense with it, updating your working memory. The test debuted in 1958 and has had a robust fan base ever since among researchers and people selling it as a working-memory booster.

The version that underlies claims that its benefits extend beyond working memory is called the dual N-Back test. As the name implies, the test taker engages in two N-Back tests at once, one of them visual and one auditory. When there's a repeat of either, the test taker has to indicate as much. As you'll see below, the dual N-Back is considered one of the two titans of working-memory enhancement. When the two are compared, we get a clash of the titans.

Lots of other tests are available on the internet that you can use to check your memory. But you should be aware of some caveats. These quizzes may not be validated—or even tested at all. Without published findings confirming an exam's validation, you could be using an instrument that exists only to collect your personal data, including your medical history. And the uses can be even more insidious. One study of tests offered online to consumers who were worried about Alzheimer's disease found that these tests were often misleading and of poor or very poor quality, even "predatory."[30] Website hosts of these tests largely failed to disclose conflicts of interest, despite many being run by organizations or companies offering products or services related to dementia (one appeared at Dr. Oz's site). The bottom line was that none of them offered information that would be useful for diagnosis, and test takers did not typically have access to post-test counseling to better understand the results. One particularly egregious example had only a single outcome, no matter what the user responded: "severe risk" of Alzheimer's disease.

Tests that clinicians use include the Mini-Mental State Examination, or MMSE, which typically is used in older patients to assess cognitive function. It assesses attention and memory, among other things. The Mini-Cog is a three-minute screener for Alzheimer's disease risk and assesses recall, not unlike the MoCA, except that it uses only three words. There are several others, all intended to screen for cognitive impairment and related conditions.

For a university-hosted test of your short-term memory, you could try one of the following:

A University of Washington site (intended for students): http://faculty.washington.edu/chudler/chmemory.html

A site begun at Harvard that includes several memory and attention tests: www.testmybrain.org/

THE SECRETS OF THE MEMORY PALACE REVEALED

You've already met the memory athletes. Now it's time to reveal their secrets.

They aren't really so secret. Many of them use an ancient technique called method of loci—sort of like having a "mind palace" where you locate items on a list. It's a technique that Greek and Roman orators, with their apparently protean memories, relied on as they orated. The association of a familiar spatial territory with items on a list aids in recalling those items in order. Instead of trying to remember an impossibly long series of items, you create a spatial story with them, locating them in your mind palace. To retrieve them in order, you just retrace that journey, retell that spatial story, and find them. Creation of stories continues to be a key tactic across all our cognitions.

As an example, for a series of five numbers (55, 89, 144, 233, 377), using your house or apartment as the "memory palace," the story begins in the bathroom, where you see 55 in the bathtub and 89 sitting on the toilet lid. You follow your usual path into the kitchen, where 144 lies in the kitchen sink, 233 sits on the sponge, and 377 looks out the window over the sink. You can create your palace from your imagination, or use a single room with specific locations where the items to remember can be stashed in order.

In that study of nonathletes who received six weeks of training, method of loci was the mnemonic approach they used.[31] When they applied it, the connections in their brains between the DMN and structures related to visual memory took on conformations much like those of the memory athletes. Rather than trying to whiteboard the information and execute a herculean hippocampal effort to maintain it there, the nonathletes went on a spatial journey, stashing each item of the series in a specific location. To recall the series, they retraced their steps and looked at where they had (mentally) left each element.

Clearly, this training improves the quantity of information you can cram into your working memory, or even into your longer-term storage, according to Dominic O'Brien. He argues that for people who want to achieve high levels of retention, deliberately refreshing the information in five steps over increasingly long time intervals can help with that.

So we have the dual N-Back test, which, regardless of its far-transfer effects or lack thereof (the latter is more likely),[32] seems to affect working memory, and we have the method of loci. One is a process you engage in and improve on over time, and the other is a strategy that requires a commitment to learning and practice. In the battle between these two working memory titans, which is better, and does either show results beyond its near effects on working memory?

One study that looked at them head to head found, perhaps unsurprisingly, that people who trained in the respective techniques . . . got better at the respective techniques.[33] We already know that for a lot of brain-improvement strategies, practice achieves some fine tuning. But do they offer any benefit beyond enabling you to show off at parties or perhaps one day compete in the World Memory Championships?

Some researchers see possible benefits for people who are experiencing memory issues because of aging or other causes, if not for unaffected people.[34] As the authors of one small study put it, "Post-training gains are within reach of older adults." I feel seen.

THE NEUROENHANCERS

Many college students are already aware of the classical ways to "neuroenhance" by way of pharmaceutical drugs and caffeine. In fact, their expectations related to pharmaceuticals may be a bit high. Studies indicate some mild improvement in memory with methylphenidate (aka Ritalin) and some improvement in attention and memory with modafinil (aka Provigil)—whereas other studies contradict those findings—but nothing leaves the user with superpowers.[35] They all have side effects, some of them potentially dire. I'm not in the business of prescribing drugs, so those issues are a topic to take up with a clinician.

Caffeine is accessible for most of us, and it does seem to give an assist with attention—although there can be a tipping point where one leans too far toward the jitters and away from concentration. That said, it doesn't require a prescription, and in a head-to-head study of caffeine against modafinil and methylphenidate, of the three, caffeine had the most obvious positive effect on sustained attention.[36] Peet's Coffee it is, then.

As wonderful as coffee is, people want pills that can fix things. Unfortunately, there aren't any that unequivocally repair whatever you think is wrong with your memory. Most drugs that have been tested don't do anything.[37] The brain is not a slot machine, it doesn't operate via a single chemical pathway, and it isn't dedicated to a single or narrow range of functions, the way, say, the heart is. Tailoring it, especially to enhance global, social, memory, and attention cognitions, and to counteract stress and anxiety, will probably require precision tools that you craft for yourself.

In the meantime, a parade of various drugs and supplements have been tried and have failed to show benefit for attention and memory.* Options on the leading edge include erythropoietin, or EPO, a hormone probably most famous for its use in doping in cycling athletes. Its primary role in our bodies is to promote red blood cell formation. But like all hormones, it exerts multitudes of different effects, depending on which tissue it encounters. Based on hints of benefit for cognition and hippocampal effects, some researchers contend that EPO could rescue memory impairment in aging or neurodegenerative disease, possibly by triggering generation of new nerve cells.[38] Testing of these hypotheses is at the beginning stages.

For you microdosing fans—or the microdosing curious—evidence suggests that if your goal is memory enhancement, you need to direct your attention elsewhere. A lot of people report anecdotally that microdosing enhances their memory and attention.[39] But those subjective experiences don't quite align with what studies show. In this case, the dose may make both the poison and the memory prop.

High-dose psilocybin does no favors for memory or attention, possibly because, well, on a high dose, the person is tripping balls and seeing All the Things as worthy of attention (possibly thanks to a backgrounded salience network).[40] What really warranted attention during the journey is something to worry about once the visions are gone and you can't figure out where you put your wallet.[41] Two tiny studies of LSD microdosing, with only twenty and twenty-four people each, found either no cognitive effects at all or negative effects on concentration. They also found a slower ability to shift attention, not because of concentration deficits but perhaps because the brain was just slower at everything. Participants in these studies, one of which used the N-Back test,[42] were not unfamiliar with recreational drugs, which might have tainted the results a tad. One loosely designed trial showed some

---

* Gingko biloba, ginseng, omega-3 fatty acids, vitamin E, and the B vitamins are among the candidates that did not keep their promises.

improvements with microdosing psilocybin, but the study sponsor was the Dutch Psychedelic Society, suggesting a bias risk.

In a gold-standard, double-blind, controlled study of forty-eight adults who weren't naïve to LSD but hadn't taken it in the last five years,* the authors found no effect on cognition across three doses of the drug, relative to placebo.[43] The one exception was that up to half the patients taking LSD reported headaches, compared with only 8.3 percent taking the placebo. In my experience, headaches do not ease a cognitive load.

Another placebo-controlled study, with twenty-four participants, took a whirl with three different microdoses of LSD. The authors found positive effects on attention and mood, slowed information processing, and increases in confusion and anxiety.[44] So if being cheerful and possibly focused (mixed evidence) while also experiencing slow information processing, anxiety, and confusion is your idea of a good time, perhaps this is the right choice for you.

Psychedelics taken in the old-school, full-dose form do have negative effects on memory, with a dose response—meaning the higher the dose, the greater the impairment.[45] No impairment is sometimes found at the lowest doses.[46] A review of the evidence up to May 2020 showed this dose-response effect on just about every memory task, whether it involved working, episodic, or semantic memory.

DO NOT JUDGE THE MACHINE

Now it's time to look at the future while we also meditate on the past. No, scratch that. While we also meditate. You've learned about neurofeedback, the process of using feedback about your brain waves to deliberately take steps to change their rhythms. Some studies suggest that it can be beneficial for working and episodic memory.[47] Authors of other studies have called it "repetitive, monotonous, and boring," which is not the best marketing slogan for a tailoring tool, effective or not.[48] And you've heard about mindfulness and meditation, the process of training yourself to dissolve that ego—unplug the DMN—and attend very deliberately to specific

---

* Is everyone except me walking around tripping all the time, making it so hard to find drug-naïve participants for these studies?

environmental cues. Now you'll hear about the two being used together to improve memory.

Researchers have found that expert meditators show increases in oscillations in the theta frequency (refresher: that's 3–7 Hz, or 3–7 cycles per second) in electrodes recording at the middle front of the head (aka the frontal midline). These oscillations are associated with the DMN, dialing down when oxygen delivery ticks up, and leveling up when delivery declines. They are also associated with the attention and executive functions, amping up when these systems take over. The ability to boost theta oscillations has been hypothesized to boost working memory (alpha waves have also been examined for their role in gating whether or not we pay attention to something).[49]

A pair of authors wondered if perhaps people who *aren't* expert meditators could, by way of neurofeedback, achieve a similar pattern of theta oscillations while using a "breath-focus" exercise.* (Focused breathing is what it sounds like: a focus on breathing deeply and deliberately for a specified period of time.†) After eight sessions, the twelve meditation-naïve participants could indeed change their theta oscillations at the frontal-midline electrodes using the same meditation strategies to focus their attention.[50] Furthermore, they could perform faster (and still correctly) on N-Back (2-Back) trials compared to their baseline performance. Twelve of their counterparts who underwent a sham neurofeedback protocol did not experience these changes.

The authors say that the DMN and the attention/on-task networks may compete with each other for resources. The task of focused breathing essentially reallocates those resources to the attention/executive function networks, giving them the upper hand. The theta oscillations that dominate during focused breathing may allow attention networks to monitor for lapses in attention and call the brain back to order—to the breathing exercise. The oscillations, they write, may be a kind of "language" that the prefrontal

---

* This is a focused-attention exercise, one form of mindfulness. Another is "open monitoring," in which you are asked to simply make note of what shows up in your brain, without latching on to it—which sounds a lot harder.

† I can't figure out how to turn off reminders for focused breathing on my Apple Watch, so sometimes I do the exercises just because the watch has asked me to.

cortex uses to tell the brain it's supposed to be focusing, thus snapping it out of mind-wandering.

In other approaches using brain-computer interfaces, researchers are taking a stab at transcranial interventions for memory. Some of what they are doing is downright spooky and futuristic, which is why you'll find it in the last chapter. But to touch on the more familiar versions, scientists have test-driven transcranial direct current stimulation (tDCS), evaluating it for, among other things, effects on memory in people who switch tasks, like shifting from looking for the oregano in its usual place (a top-down task) to dealing with a fire on the stovetop (a bottom-up task). They found no effect of the stimulation itself on outcomes.[51]

Another group assessing the effects of tDCS on working memory, based on earlier studies showing transient improvement, did find some enhancement of N-Back performance in the participants receiving the real stimulation.[52] Yet another group decided to add social stress to their testing of working memory, which is negatively affected by stress.[53] They found no effect of tDCS on working memory. So yes, results are mixed. A few studies even find benefit for some participants and negative effects for others using the same protocol.

If you remember the study from earlier in the chapter looking at neurofeedback and focused breathing, you'll recall that the feedback was predicated on what was going on with theta oscillations. Theta takeover was associated with better attention and memory. So it's probably not surprising that one tactic for improvement would be to bombard the front of the brain with theta waves created externally to mimic the pattern in the brain. This accelerated form of transcranial magnetic stimulation, called intermittent theta burst stimulation, or iTBS, relies on the notion that theta oscillations trade off with gamma oscillations as the prefrontal cortex and hippocampus get in rhythm together. The stronger this coupling, goes the idea, the better the hippocampus can hold that ink on the working-memory whiteboard.

One trial of this approach pitted it head-to-head with another form of rTMS (repetitive TMS) and with sham controls, with twenty people in each group.[54] Among the cognitive tests they used was the good old N-Back to assess working memory. The groups receiving either rTMS or iTBS showed improvement on their N-Back tasks, but the iTBS group outperformed their magnetically stimulated counterparts. They also were alone in showing significant improvement on a test of executive function. All around, the use of theta rhythms to match the brain's attentional rhythms yielded the best outcomes in this small but (at least) randomized, sham-controlled trial. Instead of using neurofeedback to condition their brains, the participants had their brains conditioned by the iTBS.

If having your brain bombarded with theta waves, which sounds vaguely space agey or new agey, depending on your perspective, is not terribly attractive, have you considered employing curiosity? It is said to kill cats, but for humans, the motivation it drives may support memory.[55] One pair of researchers at Cardiff University and the University of California, Davis, have proposed an entire framework for learning and retention that uses curiosity as a pillar.[56]

The framework begins with being wrong about something. This prediction error leads you to consider why. You appraise the wrongness or uncertainty that the error introduces. You are curious about why it was wrong or weird and what might be right, so you explore to see if you can figure it out. This process, the two authors hypothesize, gets your hippocampus and your reward circuits working together to consolidate information and distill it into something memorable. When your exploration leads to a reward that satisfies your curiosity, you've completed the cycle and made the memory.

Because this cycle involves the reward circuits, the framework, dubbed PACE (for **p**rediction errors, **a**ppraisal, **c**uriosity, and **e**xploration), could become an accessible lifetime habit (if it bears out experimentally). If you make curiosity your purpose, that's motivation. As we've seen, motivation can be the great uplifter for cognitive outcomes. If we use our attention cognition to mindfully attend to curiosity, we will almost always have a purpose. With the dopamine ping of reward we receive when our curiosity is satisfied, we'll be feeding a healthy practice. And a curiosity orientation will always give us something new to pursue in a world that continually serves up uncertainties. As a byproduct, we also might get better at handling uncertainties if we have a process in place to resolve them.

Think of some weird fact that you know. This PACE pattern may be one reason why people like weird facts and remember them so well. They pull us out of our top-down comfort, our autopilot, and snap us to attention: *Why* do echidnas make pink milk? It's a form of **p**rediction error—you'd predict milk to be white—so you **a**ppraise the situation and realize that, yes, pink milk is unusual. So you engage your **c**uriosity and **e**xplore for an answer. The uncertainty introduced by the strange trivium about echidnas' pink milk entered you into a PACE cycle. Here's another one: the little weirdos also have penises with four heads. Have fun!

As this example illustrates, we can feed our curiosity easily. We can use the same tactics we employ to bulk up our cognition, improve our social ease, and relieve stress and anxiety: pursuit of stories about echidnas . . . or about each other.

SOCIAL ENGRAMS

We need each other. The separations forced on us by the pandemic probably made that clear. But if not, I hope reading this book has made it clear. You'll be unshocked to learn, as we close out this chapter, that one of the ways we consolidate memories, revisit their meaning, buffer the stress they can cause, and PACE ourselves is through our social interactions. Approaching them with a mindful stance helps. The tailoring comes in as you fashion each of these methods to best suit you.

The memory effects of our social lives have motivated one pair of researchers at Boston University to posit that in addition to regular old engrams, we also create social engrams.[57] These memory tracks, like the general kind, rely on processing through the hippocampus, that pair of seahorses in the brain.

Uncertainty drives curiosity, and curiosity drives our memory and learning skills. That is a good explanation for why humans love gossip so much. If there's anything that lights up the brain, motivates exploration, and persists to a resolution, it's hearing some story, preferably salacious, surprising, or titillating, about another human being. I know we are supposed to deplore gossip, but it's the rare person who can turn down a chance to go into high alert and attend to a surprising story. It's even more rare to find journalists— who tell stories for a living, often telling a new one every day—who aren't also inveterate gossips.

But we also learn from each other when we share stories and have other encounters. We can mimic the behavior of others, not in a mocking way but by absorbing it and applying it in other, similar situations. To do this, we must form a social memory. With enough practice, we can eventually summon the process from top down in our autopilot mode, like starting to shake hands with someone—and then remembering social distancing. In this way, humans as a species are extremely efficient at gaining new skills without having to reinvent a process every time. But for that to work, we need other humans around us to imitate and on whom to practice our mimicry.

We can also rely on others to indicate salient information that is novel to us, which updates our database of socially relevant information. We deal with this new information from the bottom up, like other novel cues, such as a stove on fire. When you learn to use flatware from left to right at a fancy dinner, you've updated your information. A few dinners down the road, if it becomes habit, you've switched to autopilot.

Stories serve both purposes for us. They offer novel insights we can remember and use to expand our repertoire. And they give us characters we can imitate (or choose not to imitate). When we tell stories to someone, we summon memories. When we hear them from someone, we use our attention and create new memories. As we participate in these social exchanges, we execute a million social engrams, nodding at an unspoken cue, making a face, crossing or uncrossing our arms, reaching out for a hug—behaviors that most of us engage in without a thought, the product of our top-down autopilots doing their thing. These interactions in turn reinforce our brain's feelings of reward, and if they are positive they represent social bonds that may bolster our memory capacities.[58] Research has focused a lot on this effect in older people,[59] as in a study of 7,973 middle-aged or older people in China showing that increased and diversified social participation is linked to improvements in cognitive function. The people in that study were at least forty-five years old, but this diversity and appropriate dose of social participation matters to everyone.

Because it's possible to have too much of a good thing, we all have our thresholds for too much or too little social interaction, another facet of our brains that requires individual tailoring. For some of us, living with a life partner can provide just the right level of give-and-take, news of the day, novelty, support, and resonance. Most of us need at least some social interaction.

When we sit down with a friend over lunch and ask, "So, what's been going on?" and the stories start to flow, we are doing it all: offloading neuroergonomically with another person, engaging our social autopilots, paying attention, summoning memories, retaining important information on our whiteboards for follow-up, detecting prediction errors, showing curiosity, and asking questions based on what we've stored on our whiteboards. If it all goes swimmingly, we get that reward at the end, the satisfaction of resonating emotion, whether happiness, disgust, anger, or joy. We feel validated, our memories consolidated, our problems aired, our curiosity sated. And perhaps the food was good too.

CHECKLIST

✓ Engage in breathing exercises for attention
✓ Build a mind palace for expanding short-term memory
✓ Share stories, and practice attention and memory
✓ PACE yourself—make following your curiosity a constant pursuit
✓ Enjoy coffee or tea
✓ Get a dose of physical activity
✓ Give the N-Back task a try
✓ Level up to dual N-Back
✓ Engage your other cognitions—social, stress, global—to work with your attention cognition

CHAPTER 8

MOOD COGNITION: MANAGING MELANCHOLY

All things are full of weariness; a man cannot utter it.

Ecclesiastes 1:8

IN ABOUT 900 CE, Ishaq Ibn Imran wrote the oldest known surviving text focused on depression, or melancholia. He was an Iraqi medical man living in Kairouan, Tunisia, who had studied the work of his famed Western peers, including Galen and Hippocrates. He was ahead of his time in other ways (or perhaps later times fell behind), believing that a balanced diet and physical fitness were key aspects of treating diseases. Today, we would call these "modifiable lifestyle factors." For people experiencing mania, for example, he recommended cold, moist foods (which is not a current recommendation).[1]

Here is how Ibn Imran defined melancholy: "Melancholy affects the soul through fear and sadness—the worst thing that can befall it. Sadness is defined by the loss of what one loves; fear is the expectation of misfortune."

He also had some correct ideas about the forces that operate to result in depression, even if his name for them differed from what we'd call them today. He figured that people could inherit a tendency to it, although his mechanism a thousand years before anyone knew what a gene was relied on problems with sperm and the uterus. He also believed that six factors in the environment might act to cause depression.* Ibn Sina (Avicenna to Westerners), the more famous Persian man of science before science had a name, popularized these ideas sufficiently in the tenth century that they lasted several hundred years.

In the seventeenth century, a man named Robert Burton, writing as Democritus Junior,† took up the banner of treatises on melancholy.[2] An anatomist, he wrote a book, *The Anatomy of Melancholy*‡ (1621), in which he attempted to tie together the roots and structures in the human body that might be responsible for depression. We've gotten a bit better at understanding these things with our fancy imaging tools and so on. Burton himself applied one of the therapeutic measures we use to this day: pursuing a purpose. He sought to ease his own melancholy by offloading what he called the "pus" in his head.

That "pus" in our heads, aka depression, affects almost fifteen million people in the United States and more than three hundred million people globally, some of whom may not even realize that's what the "pus" is. Like all things related to the brain, mood exists on a spectrum. We can find ourselves moving around that spectrum and landing for only brief periods or, unfortunately for some of us, for years or even for a lifetime. People associate the mood condition of depression with behaviors like frequent crying or withdrawing from people, but it can also manifest as irritability, angry outbursts, working constantly, sleep troubles, substance abuse, or even back pain.[3] When things have gotten that intense and lasted for

---

* Ibn Imran met an unfortunate end. He was summoned to advise a sultan who had some kind of mental illness. Thanks to a rivalry between Ibn Imran and another physician that left the sultan suspicious of Ibn Imran, the ruler had him executed.

† Democritus, the "Laughing Philosopher," was supposed to have received a visit from Hippocrates while studying under a tree in his garden, so in its way Burton's nom de plume is a straight line from one clinician to another.

‡ The amazing full title is *The Anatomy of Melancholy, What It Is: With all the Kinds, Causes, Symptomes, Prognostickes, and Several Cures of It. In Three Maine Partitions with Their Several Sections, Members, and Subsections. Philosophically, Medicinally, Historically, Opened and Cut Up.*

some time, the "mood" has transitioned to a "mood disorder" known as depression.

This chapter does not contain medical advice. MDD, or major depressive disorder, is a clinical diagnosis that requires professional treatment. If you think you have depression or are concerned that you are headed that way, please consult a mental health expert or contact a helpline for assistance. In the United States, the government maintains a helpline at 800-662-4357.

Depression is not, of course, the only mood disorder. We can experience moods that shift often, even quite rapidly, leaving us feeling mercurial and unpredictable, even to ourselves. These swings can be relatively low in amplitude, or they can be big and obvious. As with depression, if you feel that your mood or mood changes are affecting your daily life and relationships, please consult a mental health professional.

Let's think again, or ruminate, on rumination. If knowledge is power, we can use that power to bolster our mood cognition. Rumination in a negative sense is when you experience highly self-critical thoughts[4] that reinforce negative feelings and a sense of doom, distress, or fatalism. They intrude when you're washing dishes, when you're driving, when you're trying to sleep, when you're supposed to be having a good time. Engaging in such thoughts requires you to focus on the self, review autobiographical memory (how else can you flog your past self and make your present self feel terrible?), and contemplate how terrible you are at doing the thing you're trying to do right now.

These processes signal to you that your DMN is active—jogging your self-referential, autobiographical memory—at the same time that your task-related networks are calling attention to your self-perceived awfulness in the present context. Your hippocampus obviously plays a role, serving up those autobiographical memories with a dark tinge, thanks to the amygdala. And despite what the people who think positive thinking is a cure-all for everything will claim, you just cannot make the cycle stop. It draws a curtain over everything.

When this sequence leaves you feeling numb and unable to experience joy or any other emotion, you still feel something: anhedonia, which means a loss of *hedone*, or pleasure. You know it's there somewhere, but you can't feel it. When rodents are in this state, they are less inclined to consume sweetened water when it is offered.[5] When we experience it, we are unable to enjoy the sweeter things in life.

Why do our brains allow this intense level of rumination? As with anxiety, there is some utility to it.[6] If we didn't ruminate by revisiting and reexamining our experiences and conducting autobiographical reviews, we might never learn from or correct our behaviors. In that sense, rumination is a form of reappraisal, and when we wield our reappraisal tools in a skillful way, they help us. In a social species, being able to do that is adaptive and important to survival, just as anxiety can be protective when it's appropriate for a situation. Indeed, the worry we associate with the feeling of anxiety is itself a ruminative process: we are considering ways to solve problems that have yet to arise but could, maybe, possibly do so.

Dealing with mood derailments is crucial. In earlier chapters, we drew direct links between health and whatever it is that IQ tests really measure, and between health and social cognition, anxiety and stress, and memory and attention. But unaddressed mood dysregulation, especially if it involves maladaptive rumination powerful enough to turn into depression, threatens everything: our cardiovascular health, our relationships, our immune responses, and for the estimated one in five people with depression who attempt suicide at some point, our lives. And it interacts with all other aspects of our brain's function, drawing the curtain on cognition, social or otherwise, dancing in lockstep with anxiety, clouding memory, and distracting attention. Its reach is long, its depth can be unfathomable. So what can we do if we feel its cold touch or find ourselves drawn into its abyss?

If you have serious concerns about your mood, again, a mental health professional is what you need. If you feel that you're sliding into a negative state and you'd like to arrest the momentum, you can start by taking some tests online to see if your perception is correctly assessing your status. The caveat, of course, is that not all online tests exist out of pure altruism, and you might want to check the fine print regarding what the website hosts do with your data. Just below, I've listed some assessments hosted on government or university sites. You can also check out the list offered by the American Psychology Association: www.apa.org/depression-guideline/assessment.

In mood-related studies, researchers tend to rely on a fixed set of screening tests and scales. Common ones you'll see include the Beck Depression Inventory (sometimes with II), or BDI, for people ages thirteen to eighty. With twenty-one items to answer, it takes about ten minutes. Another is the Hamilton Rating Scale for Depression, or HRS-D (also called the HAM-D because no one in these fields can agree on using a single term for anything), which tracks symptoms of depression using twenty-one items. (There is also the HRS-A, for anxiety.) The scores at baseline, during treatment, and posttreatment can be used to track the effects of treatment, so some studies use this tool to determine the effects of a candidate intervention versus a placebo or sham.

Another scale that captures changes through time is the Montgomery-Åsberg Depression Rating Scale, or MADRS. This one is also accessible to anyone, and you can find it here: www.apa.org/depression-guideline /montgomery-asberg-scale.pdf. As we look at the evidence for what does and does not seem to work for mood, you'll see these scales used often.

The UK National Health Service offers an online depression and anxiety self-assessment quiz here: www.nhs.uk/mental-health/self-help /guides-tools-and-activities/depression-anxiety-self-assessment-quiz/.

The University of Texas-Southwestern Medical Center's O'Donnell Brain Institute offers an online self-test for depression here: www.utsouthwestern .edu/education/medical-school/departments/psychiatry/research/center/self -rating-test.html.

You can find the Quick Inventory of Depressive Symptomatology-16, or QIDS-SR16, a self-report instrument that is often used in research, on UCLA's website, here: http://narr.bmap.ucla.edu/docs/QIDS-SREnglish16item.pdf.

A quick look at these questionnaires gives you an idea of some of the persistent signals that a person's mood is flattening or declining, including changes in sleep and appetite, body weight, ability to concentrate or focus, energy levels, suicidal ideation, and anhedonia. The tools available to tailor your mood are highly individualistic, and just as with pharmaceutical and behavioral therapies, what works for one person may be a dud for another. The key, as Dory put it in *Finding Nemo*, is to keep swimming.*

---

* Yes, I have watched a LOT of Pixar movies.

THE MAGIC OF MOVEMENT

Or running, or walking, or doing tai chi. As you may have noticed by now, some form of movement or physical exercise, if you are able to do it, is the all-purpose tool in our tailoring kit. Evidence shows that it mitigates stress and anxiety, affords benefit for memory and attention, and can truly uplift cognition, including social cognition. With those effects, especially on stress and anxiety, the drivers of our brain's moodmobile, it should be no surprise by this point to learn that, yes, exercise can sand down the edges of a rough mood.[7]

I know that for some people, constraints of time, disability, wherewithal, and resources affect access to physical activity. But it's important to recognize when another kind of constraint is acting on you, the very one you want to ameliorate. As with anxiety, depression and its kin on the mood spectrum can get in the way of your getting going.

Why is moving around so magical? It's not magic. It's molecules. When *we* move, *they* move to our brain more easily, especially the most crucial molecule of all: oxygen. A hard-core workout can trigger mood-lifting endorphins. And over time, sustained daily or near-daily exercise releases molecules, such as brain-derived neurotrophic factor, or BDNF,* that can prompt increased production of neurons, which in turn can make fresh connections.[8]

Whether you get high on the pain of a Peloton workout or take fifteen minutes to walk around the block or your couch, just about any physical exercise is better than none at all. But being active for at least twenty to thirty minutes, three to five days a week, regularly, is better than sporadic bouts. You don't have to be an ultrarunner to gain benefits.

If you think about physical activity as a therapy or a dose of medication—well, a lot of us commit to taking some kind of medication at a specific dose, every day, often for a lifetime. Physical exercise is like that: we have to start it, and we have to maintain it. Canadian guidelines for the use of "complementary and alternative medicine," or CAM, even recommend exercise as a first-line single therapy for people who have mild to moderate depression.[9]

---

* The name indicates that it originates in the brain (brain-derived) and encourages growth (-trophic) of neurons (neuro-).

Exercising involves interactions with environmental factors, including social factors, that probably play into its effects on mood. Because of its beneficial impact on the immune system and as an anti-inflammatory agent, exercise can support you in a fairly robust defense against disease.[10] Often it gets you outdoors, and the benefits of being outside are immense for a human brain that, after all, evolved outdoors. If you walk or work out with friends or even with a perky Peloton coach cheering you on, you're leveraging the profits of being social as a member of a social species.

I have a friend who's written an excellent book about exercise and recovery. She herself is an athlete who skis or mountain bikes every day. Clearly, she gets enough exercise. Yet one of the rules she and her partner follow is that no matter what else happens that day, their evening walk is nonnegotiable. That isn't only about the physical activity. It is also about the social facet, the offloading of cognitive burden, the sharing of a day. Exercise is never *just* exercise.

THE PILL QUESTION

If you're looking for a real medication to take, in addition to or instead of your dose of physical activity, these options require as much of a tailored approach as exercise does. Most antidepressants rely on the idea that you're low on a signaling molecule in the brain that regulates mood, such as one of the neurotransmitters serotonin, dopamine, and norepinephrine, and the job of the pills is to boost levels of that signal, one way or another. Three pathways are available: (1) be the signal, that is, mimic the actual molecule in some way; (2) block breakdown or loss of the actual molecule, that is, make it last longer so that its signal does too; or (3) keep it from signaling if its usual role is to tamp down mood. These assumptions are controversial and remain much debated.

One reason for ongoing debates is that for about one in two people with depression, prescription drugs are effective.[11] Averages suggest an effect only slightly greater than placebo, but that reflects almost a binary: the effect seems quite large for some and nonexistent for others. A wide-ranging Cochrane review, which is considered fairly rigorous, showed that these interventions, among several categories the authors analyzed, offer benefit for some people some of the time, but also carry a risk for "significant harm."[12]

The dispensation of pharmaceuticals remains a job for professionals. What nonprofessionals in these modern times want to know is, What about the psychedelics? When I say that psychedelics are experiencing a renaissance, it's not an exaggeration. Interest in them has grown considerably in the twenty-first century, with more than a dozen clinical trials conducted in 2020 and another dozen in 2021.[13] The good news is that if the trials involve genuine scientific research, rather than just, "Hey, man, microdosing really turned me around," and if they are placebo controlled, we can get a good idea of whether or not these substances truly affect mood.

So far, what evidence there is points to a positive average influence that's directly relevant to mood: reduced focus on negative stimuli, decreased rumination, increased empathy, lower sensitivity to rejection, less tendency to withdraw socially, and, relatedly, "ego dissolution," a breaking of the egocentric circuit to allow connection with others.

In another similarity with prescription antidepressants, the drugs used for microdosing also tend to act by one of the three pathways described above that affect mood-regulating molecules. They can act through one subset of proteins that recognize serotonin, a key mood-regulating signaling molecule, and possibly imitate its effects.[14] The best-known and most commonly studied psychedelics in mood-mitigating research are LSD and psilocybin, the former synthesized in a lab and the latter lifted from a fungus.

Here's what we know so far. Open-label studies of psilocybin suggest benefit, but by the very nature of open-label trials, the participants knew what they were taking. At any rate, in one such study, twenty people who took psilocybin once felt less anhedonia at one week, and they felt more social and had higher openness scores even three months after their treatment.[15] They also reported that with psilocybin, they felt more encouraged about connecting with others, whereas with their previous treatments—either short-term talk therapy or prescription medication—they had felt more disconnected from and avoidant of other people. That's not a head-to-head comparison, just a retrospective, self-reported impression. The results of this small study suggested that psilocybin might have lifted mood by reinforcing social cognition, which in turn reinforces just how important social cognition can be.[16]

In what probably sounds like an odd head-to-head comparison, another set of researchers compared the effects of a dose of psilocybin to a dose of niacin (a B vitamin), both accompanied by psychotherapy.[17] Their patient group of fifteen was highly selected, all of them having received a psychiatric diagnosis related to their having cancer. The reason they used niacin as the "placebo" is because at a sufficient dose, this compound can produce a flush and a rush, which serves as a match to what someone might expect from a low dose of psilocybin. In this trial, the small number of participants had psilocybin at a first session and niacin at a second, or the reverse, so inducing some kind of feeling with placebo was important to truly mask the substance's content.

The upshot was that at six months, up to 80 percent had experienced what was gauged as a significant antidepressant or antianxiety response from baseline, which persisted to 4.5 years. They "overwhelmingly" reported having had positive life changes because of the psilocybin add-on to their therapy, rating it as "among the most personally meaningful and spiritually significant experiences of their lives." The authors called the magnitude of reduction in depression and anxiety symptoms "large," but also emphasized that the drug was an addition to psychotherapy.

The relevance of having professional psychological support for people trying psychedelics is a common theme. A 2017 study, also open label, evaluated the effects on depression of psilocybin administered "in a supportive setting," and found that the twenty patients showed "marked reductions" in their QIDS-SR16 scores in the first five weeks after treatment.[18] In fact, four of them were in remission—that is, showed no signs at all. The effects held up at six months.

Some of the work of this lab and others implicates altered DMN connectivity in the before-and-after effects of psychedelics.[19] But the DMN is not the only circuit of interest. When our mood plagues us, we likely want the DMN to stand down, sure, but our executive network needs to perk up a bit too. Some research suggests that the brain goes through a reset of these networks, including their connections with the amygdala and hippocampus, in the short term after psilocybin exposure, before settling into connections that yield a healthier mood state over the course of a few weeks. People in these trials even report feeling refreshed,[20] like Infrastructure Week for the brain.

Another hallucinogen, salvinorin, instead of mimicking serotonin, acts on cells through the same proteins that some opioids use. But it relies on a different set of proteins from the one associated with opioid addiction: a signaling route that doesn't interact with addiction circuits. Another feature of salvinorin, which is derived from the plant *Salvia divinorum*, is that it hits like a truck.

Brain scans of study participants show that within minutes of being inhaled—the typical administration route—salvinorin breaks up the strong connectivity of the DMN and leaves the brain with that far-out feeling of oneness with the universe, instead of separation from it.[21] In other words, its effects are quite similar to what other studies show with psilocybin and LSD, but a lot faster. They also subside pretty quickly, or at least the hallucination part of the experience does, within about fifteen minutes.

All of which is to say, you shouldn't take what you read here as sanction to go off on your own and light up a giant salvia spliff. Again and again, studies emphasize the importance of an expert and therapeutic human presence.

Another reason not to reach for the salvia or the mushrooms or the LSD right away is the usual: these studies often suffer from the same drawbacks as so many studies in neuroscience. They have small numbers of participants, focus on different brain networks or only one at a time, use a vast array of methods and designs, and often lack any controls or at least placebo controls.

For example, a randomized trial of psilocybin in patients with depression used wait-listing as a control, meaning one group of participants waited eight weeks for treatment while the psilocybin group underwent their intervention. That's not optimal for a lot of reasons, and the study was small, with only twenty-seven people. But the results at least were not "mixed" relative to other psilocybin studies. The trajectory remains the same: improvements for the psilocybin group, in this case on HRS-D scores, which the authors say decreased rapidly and more substantially than in the delayed-treatment group. Yes, the people waiting for treatment also experienced some improvement, which is not uncommon.[22]

How does psilocybin stack up to antidepressants? One of the key figures in this research field, Robin Carhart-Harris at Imperial College London, and his team just completed a study examining that question, sort of. The trial involved fifty-nine participants and pitted psilocybin (25 milligrams in each of two doses taken three weeks apart) plus daily placebo against a microdose of psilocybin (1 milligram in each of two doses taken three weeks apart) combined with a daily dose of the selective serotonin reuptake inhibitor escitalopram. Neither intervention bested the other for improvements in QIDS-SR16 scores at six weeks of treatment, with only a two-point gap between them in score declines.[23] In raw numbers, the higher dose of psilocybin looked more promising, with 70 percent experiencing a positive response to it compared with 48 percent of those taking the microdose plus escitalopram. But the scatter among the scores was so wide that the two groups did not differ significantly, even with this 22 percentage-point gap.

A gap of 28 percentage points separated those in the higher-dose psilocybin group who experienced remission from those taking escitalopram plus microdoses. Several other comparisons, including of scores on other depression scales and measures of anhedonia and "flourishing," suggested an effectiveness edge for psilocybin plus placebo, but the way the analyses were performed means these "differences" could reflect outcomes by chance rather than because of a real effect.

The trial was not large, and some other factors point to a need for bigger studies. The people in the higher-dose psilocybin group had experienced depression much longer, at twenty-two years on average, than the fifteen-year average in the escitalopram group. Only about a third of the participants were women, most were White, and the study was so small that looking for any sex-based effects likely would have been futile. The authors also point out that antidepressants like escitalopram can take longer than the six-week duration of the trial to produce a benefit. This delay is common to these drugs, so it's unclear why they decided on six weeks as the trial duration.

The authors also mention that when trial candidates signed up, they more often evinced interest in psilocybin over escitalopram, suggesting a skew in the trial population toward people interested in psychedelics. The psilocybin was the draw for this small study, highlighting the continued attraction and popularity of these intervention modes. Given that, one bit of good news tied to this lead balloon is that people in the higher-dose psilocybin/placebo group had no more adverse events than those taking two tiny doses of psilocybin and daily escitalopram.

While the world continues its wait for bigger studies and more data on the psilocybin front, the authors of the earlier wait-listing study were keen to point out that the effects of psilocybin seemed to outlast those of ketamine. That observation suggests that if there's a drug to beat in the psychedelics-as-therapy competition, ketamine may be the one.

THE KETAMINE QUESTION

Ketamine has been many things to many people since it debuted in the 1960s. It was first used as an anesthetic, attractive on the battlefield for its rapid action and then in veterinary practice. Later in the century, it gained popularity as a "club drug," a perfect match (for some people anyway) for hypnotic, repetitive dance music. As with its counterparts in the loose society of mind-altering drugs, ketamine binds a protein in brain cells that normally relays messages into these cells. In this case, ketamine blocks the messages; it's an inhibitor of the signaling.

The proteins targeted by ketamine normally bind glutamate, the excitatory signal for neurons. In the presence of ketamine, the neurons don't get stimulated. That doesn't sound very antidepressant-like, but there's a plot twist. The target cells have a job, and that job is to inhibit other cells from releasing glutamate. To do this, they have to be excited themselves. It's like waking up security to shut down a rowdy party. With ketamine around, these cells aren't excited, so they don't wake up and inhibit other neurons. The net effect is that those other neurons stay perky and invigorated, which does sound more like an antidepressant and probably explains the feeling of a rush and a high that ketamine gives.

Ketamine exists in two forms, each a mirror image of the other, like holding your left and right hands palm to palm. They are designated as "R-ketamine" and "S-ketamine." The one that's used as an antidepressant is the S form, dubbed "esketamine." This drug has been approved by the US Food and Drug Administration as a nasal spray (trade name Spravato) for people whose depression has not responded to other treatments. It's remarkably effective for many people, which raises the question of why they have to work their way through so many other drugs to get to it.

Curious about this treatment, I asked Carlene MacMillan, a Harvard-trained psychiatrist who runs a pair of clinics in New York City, to tell me about the experiences of her patients. She noted that the FDA has approved the drug under a strict protocol called the Risk Evaluation and Mitigation Strategies, or REMS, program. If you've ever taken the acne drug Accutane, you're probably familiar with the REMS process. Patients have to sign up for a registry, and the drug itself must be delivered to the provider's office—it can't go straight to the patient.

Once the esketamine is available, the patient has to set aside at least two hours, the first ten or fifteen minutes of which is spent preparing and using the spray. MacMillan said that the patients' blood pressure must be monitored, because esketamine can cause spikes. "Most people will experience a very transient increase in blood pressure, but not massively so," she said. Then the patient kicks back and relaxes for a couple of hours, in part because the REMS protocol requires it. "Even if they feel ready to leave before the two hours is up, they need to stay," MacMillan explained, saying that most people feel ready within an hour. In her office, the patient can spend that time listening to music or undergoing guided meditation, whatever they choose. Afterward, patients "generally go on with their day, feeling enough back to normal that they are able to function," she said. Most don't experience dissociation during the treatment. "The dosing is low enough that those effects are pretty mild." Dosing and frequency are a decision to be made with a clinical professional, but patients might start with two sessions a week, taper to once a week after about a month, and then to once every two weeks.

And how well does it work? Response rates vary, MacMillan said, but esketamine bests traditional antidepressants. It may do so by converging on the same connectivity effects as traditional pharmaceuticals, with the executive network regaining the upper hand.[24]

Ketamine can also be given as an infusion,[25] but any psychiatric use of infused ketamine is off label, MacMillan said. Nevertheless, infused ketamine has been studied, and it offers what one group of authors described as "rapid, profound, and surprisingly durable antidepressant effects," to their "amazement."[26] They found that ketamine infusion was effective in patients with treatment-resistant depression, including some who had bipolar disorder, and that a third of patients experienced remission. They noted that the dose that worked varied from patient to patient, as with most such interventions.

Studies in animals suggest that one effect of ketamine is to promote the rapid development of sites on the neuron that receive inputs from other neurons. These sites, the spines that bloom on the "fright wig" dendrites sticking out from the cell, are the locations of synaptic activity. When cells develop more spines, they are showing a flexibility to expand. Neuroscientists call this phenomenon "plasticity." This blooming of spines after ketamine exposure suggests that the brain is exploring new pathways and perhaps dispensing with old ruts.[27]

A long-standing intervention for depression that refused to yield to any clinical treatment was electroconvulsive therapy, or ECT, which is controversial for a lot of reasons. It sounds violent; in the old days, it was. Today, someone undergoing ECT is sedated and remains unaware of the experience, and the shock that is delivered is rapid and, evidence suggests, extremely effective.[28] But it also is linked to memory loss, especially of more recent memories, and researchers have sought ways to minimize this obviously unwanted side effect. One tactic has been to deliver the shock to a single side of the brain only, which seems to carry less risk of memory loss than bilateral delivery.

Some studies suggest that ketamine could be a candidate replacement for ECT, having once even been reserved for people who could not undergo ECT for whatever reason. Ketamine, however, carries a risk for abuse, because it can be addictive. That can be managed with a careful approach.

THE SUPPLEMENT SETBACK

The interventions I discussed above require an intermediary in the form of a clinician or researcher (unless, of course, you choose to try these drugs in street form, which I do not endorse). More accessible candidate treatments exist, but unfortunately their odds of winning the efficacy race are not good.

I was really rooting for omega-3 fatty acids—sourced, for example, from cold-water fatty fish such as salmon—but the evidence, as it so often is, is mixed at best. Observational studies, which lack controls, were promising, suggesting some benefit of omega-3s for mood, making these fish-derived fats into superstars. Even some imaging studies pointed to a link between low levels of omega-3 and changes in brain structure.[29] The correlation makes sense to some extent because the brain needs omega-3 fatty acids to be a brain, and the only way humans can get these fats is to eat them. But as with so many situations in which scientists thought, Oh, A does B in the brain, so people can just take more A and get B . . . it doesn't quite work that way.

One reason is probably that the "A" in these cases is a component of our diet. Singling out a single dietary substance, in this case omega-3 fatty acids, in supplement form removes it from a broader nutritional context that may be relevant in some way to its benefits. It's not like researchers haven't thought of that, and observational studies also have suggested that diets high in omega-3s confer a benefit to mood. That led them to see if levels of omega-3s in the blood could be linked to mood effects. Some studies said yes to that question. Others said no.

When omega-3 supplements are entered into randomized, controlled trials, the superstar status granted them by observational studies starts to lose its sparkle. One meta-analysis of four randomized, controlled trials with a total of 153 children and adolescents found no effect of supplementation on depression scores.[30] As a contrast, a meta-analysis of twenty-six double-blind, randomized, placebo-controlled trials that included 2,160 adults found an effect of one kind of omega-3 (eicosapentaenoic acid, or EPA) on decreasing depression symptoms.[31]

But that's the only analysis with positive results to put the mix in mixed. Another meta-analysis of thirty-one trials that included a total of 41,470 people showed "little or no effect" of supplementation on depression or anxiety symptoms. In fact, these authors found a slight increase in risk for depression symptoms with one specific kind of omega-3 (called linolenic acid), although 1,000 people would have needed to take it for one person to experience this risk.[32]

Just to hammer that nail flush, yet another clinical trial assessing omega-3 supplementation in people with bipolar disorder found no effect on number of mood relapses, hospital admissions, or medication adjustment compared with placebo.[33] Ditto with a trial of omega-3s for postpartum depression.[34] A large European trial of multisupplementation with omega-3s in 1,025 people with body mass indexes greater than 24.99 and high scores in tests of depression symptoms showed no preventive effect against a depression diagnosis or benefit in symptom reduction.[35] The researchers also measured levels of omega-3s in blood samples from the participants and found no link between increased omega-3 levels because of the supplementation and changes in depression symptom scores. One research group tried adding omega-3s to sertraline, a typical antidepressant, for 155 patients diagnosed with depression who also had high risk for coronary heart disease, and they found no effect at all on depressive symptoms compared to adding regular corn oil instead.[36] (Yes, I wondered too, and no, corn oil does not contain omega-3s.)

The only good news is that taking omega-3s probably doesn't carry a huge risk of harm. The FDA has approved prescription versions for people who have high blood levels of a kind of fat called triglycerides, which is not desirable, and says taking 3 grams a day or less of omega-3s is probably safe.

I have the same bad news for vitamin D, another molecule initially granted superstar status. A huge randomized trial of 18,353 adults age fifty or older with no depression symptoms at entry found no effect of vitamin D3 supplementation on their chances of developing such symptoms during the median 5.3 years of the trial.[37] To quote the authors, who felt compelled to say this three times in their report, "The findings do not support a role for supplemental vitamin D3 in depression prevention among adults."

The failure of vitamin D and omega-3s in these studies causes a bit of a problem for people who feel that "nutritional psychiatry"[38] holds promise for mood and other aspects of our brains we might want to tweak. One dietary approach that might offer benefit is the "keto diet," which is peddled to consumers in many untested forms. To be clear, this is not a diet I would personally try, because the benefits are ambiguous to me, and it is not recommended for people who have concerns about cardiovascular risk.

At my age, I have such concerns. That said, some groups of researchers have hypothesized that the benefits of keto for epilepsy might transfer to a benefit for mood.[39] The keto diet for epilepsy is quite specific, carefully designed, and clinically proven, meaning it's not as easy as eliminating bread and other carb-dense foods and hoping for the best.

Some researchers figured that a keto diet with an energy intake of four parts fat to one part nonfat (or, stated another way, a fat:nonfat ratio of 4:1), which is what controls seizures for people with intractable epilepsy, might work for depression because of some similarities between the two conditions. Among them is the effect of the keto diet on GABA-related transmission. GABA is, as you may recall, the inhibitory yin to glutamate's excitatory yang. Bumping GABA signaling means less excitability in the brain, and that is, of course, meaningful in epilepsy because seizures are the manifestation of hyperexcitability. GABA derailments have been reported to track with changes in mood, and some drugs used for anxiety and depression, including selective serotonin reuptake inhibitors, trigger GABA boosts.

Perhaps if the diet route is the way you'd like to go, the Mediterranean diet, with its emphasis on "healthy" oils, would be easier and, let's face it, more fun to try than a strict keto diet. This eating approach emphasizes intake of fruits and vegetables, whole grains, and healthy fats with moderate amounts of dairy and limited red meat. One reason this way of eating is more fun, though, is that it adds the components of sharing meals with family and friends and being physically active, both of which also offer benefits for all of our cognitions. Some studies have associated adherence to this eating plan with improvement in depression symptoms.[40] At any rate, it's linked to a host of other health benefits and doesn't carry the lipid-related cardiovascular risks that can be associated with keto.

When it comes to tailoring your brain, these considerations are important. Your brain is not a clinical trial, and the risks of any intervention you undertake should always be as minimal as possible compared with its benefits. This is another reason that what works for someone else might not work for you. We all have different entry points for risk. People with depression that hasn't responded to medication after medication live with a risk level from their condition that likely dominates the hazards presented by esketamine or the keto diet. If the risk of continuing as you are is greater than the risk of trying an intervention, that's your individual calculus.

DEPRESSION AND THE DEPARTMENT OF DEFENSE

You may have noticed that we learn a lot from people with epilepsy when it comes to what we know about the brain. The location of seizure activity and the symptoms produced by seizures inform us about the role of that brain region. And interventions we use for seizures, such as the keto diet, point the way to their use in conditions that show overlap with epilepsy.

To be more specific, the interventions allow us a peek into the brain in ways that we've never been able to access before.

The US Department of Defense is very interested in your brain. It funds all kinds of initiatives related to it, with the aim of sussing out ways to address mental health for service members, gain clarity of thought and speed of reaction in battle, and connect the human brain with computers so that thoughts instead of bodies can guide (perhaps robot?) boots on the ground.

The corner of the DoD that runs this show is called DARPA, which stands for Defense Advanced Research Projects Agency (its name began without the "D" back when Eisenhower established the agency in 1958, but eventually the powers that be gave up and made the connection obvious). DARPA has devised a number of brain initiatives that have yielded information that might benefit regular people, in addition to service members. Among these is the SUBNETS, or Systems-Based Neurotechnology for Emerging Therapies, program, which targets treatments for neuropsychiatric conditions.

In part, DARPA researchers sought to capture what happens across the brain when mood changes. But mood changes aren't necessarily predictable, and to really capture them in sufficient detail, the measurements would need to be made from inside the brain. It's not ethical to poke electrodes into people's brains just to capture their mood states. For people whose epilepsy doesn't respond to medications, however, devices can be implanted because of a clinical need to monitor their brains for the origin of seizures.

Thanks to a group of volunteers who had such implants, investigators from the University of California, San Francisco, Massachusetts General

Hospital, and University of Southern California were able to ascertain the patterns associated with mood. Working under DARPA funding, the researchers asked the volunteers to track their mood states on a questionnaire for several days, while the electrodes were picking up activity in their brains. The data revealed patterns of signals that stood out with specific mood reports.[41]

The researchers used these patterns to develop what they called a "decoder," a system that can recognize signals and predict mood. When they tried the decoder, they found that it predicted pretty well. The strongest signature involved activity in the amygdala, our emotion gate, and the hippocampus, our memory gate. As you've probably noticed, this experiment has shades of the movie (and the Philip K. Dick short story) *Minority Report*, in which the government knows what people are going to do even before the people themselves know.

The UCSF researchers also tried to stimulate a region of the brain implicated in depression, the orbitofrontal cortex, which lies within the prefrontal cortex, just over the eyeballs. The stimulees, who in addition to epilepsy all had moderate to severe depression when the study began, experienced improvements, more so with a greater "dose" of the stimulus. The investigators figured that the effects in this region were communicated to other parts of networks involved in processing emotions, yielding related improvements.

The implanted electrodes measure brain waves, those alpha, theta, and gamma waves that pass back and forth, or oscillate, across brain regions as neurons take up a collective signaling, like a crowd in a sports stadium alternately standing and sitting, then reversing the wave and sending it in the other direction. We can't implant DIY electrodes in our brains and pick up our own neural signatures of mood. But using neurofeedback, someone can put them on the outside of our heads, and we can see what we can do to control the waves.

The neurofeedback process does show promise for mood. A meta-analysis of twenty-four studies that included a total of 674 people with diagnosed depression, 194 of them as controls, found improvement with neurofeedback, but the researchers didn't seem very impressed. The studies they looked at were not in compliance with "the most stringent" approaches to design and conduct. But the authors noted with optimism that studies seemed to have improved over time.[42] They're not wrong. The first such study, conducted in 1992, had only eight people with depression, all on medication, and eight without, unblinded, and the results were disappointing.

In their report, these authors called the patients the "protagonists of their treatment," which is a nice turn of phrase. Unlike "brain training" that has you playing memory or reaction-time games with the hope that by perfecting such skills, you'll generalize beyond them, you are using your brain to detect patterns of oscillations and change them, directly influencing your brain waves based on specific, usually visual, cues. Eventually, you can cut out the middleman—the computer that's giving you the feedback (cues) about your brain waves—because you will have conditioned yourself to achieve this fine-tuning at will. That's the idea.

In the end, the authors of the meta-analysis figured that better-designed and -controlled studies would have to be conducted before the hints at mood symptom improvement with neurofeedback could be taken as real, and for the right targets to be confirmed. They also observed that neurofeedback might be best investigated as an add-on to existing therapies for people with depression. At any rate, the risk of harm is said to be low.

WHY CAN'T WE START WITH THIS FOR DEPRESSION?

You'll recall TMS as another intracranial stimulation tool, one that doesn't require the patient to be the protagonist. In this case, the patient is a passive recipient. Like esketamine, TMS devices have gotten the nod from the FDA and have been certified by European regulators for use in people whose depression refuses to respond to treatment. MacMillan, who offers both esketamine and TMS, said that people usually come to her clinic to ask about esketamine, which gets more attention on the public stage.

Because the two therapies are intended for a similar patient population, with some differences in the types of treatments that have "failed," the decision about which one to choose can come down to certain factors. Risk for substance use or a problem with blood pressure might argue against esketamine, and a seizure disorder would argue against TMS. Sometimes, in the United States, the final choice is based on what insurance will cover.

Anecdotally, what MacMillan sees in her practice is that TMS yields bigger effects, comparable to those produced by ECT, than esketamine, and that the effects last longer. But what she likes about esketamine is its rapid result, making it useful in the immediate term for patients who, for example, are suicidal. "Just shifting that perspective and reminding the person that this isn't a state of being, it isn't a permanent situation" is an important lesson that treatment with esketamine can teach, she said. "It can feel like an impossible boulder to move, and this allows some kind of shifting to happen."

TMS is painless and is associated with almost no side effects. Each treatment session takes about twenty minutes, and a typical course involves thirty-six sessions in a couple of months. That means spending a half hour at the clinic five days a week for about six weeks, a time commitment that can be a barrier to receiving TMS as a therapy. The procedure starts with finding the motor cortex as a point on the brain's map. The patient wears a helmet that sends a stimulus; a resulting finger twitch indicates that the technician has found the correct position of the helmet. For the real session, the helmet will be positioned to stimulate a region in the frontal lobe, the dorsolateral prefrontal cortex, which is a target for depression.[43]

The technician finds the lowest level of stimulus that gets a reaction in order to use the minimum amount of power and avoid the risk of triggering a seizure. They check this at every session because the brain's threshold for stimulus changes all the time, thanks to sleep, diet, and medications. Because of this variability, MacMillan said that if the stimulus threshold falls significantly for a patient from one session to the next, the clinic may postpone the session. "We send them home for the day, because for some reason, the brain is way more excitable that day." The treatment stimulus itself starts at 120 percent of the lowest threshold that triggered a finger twitch.

The patient has a little bit of work to do while they're under the helmet. "They are supposed to be thinking of and engaging in things that generate positive thoughts," MacMillan said. "Not scrolling through social media and looking at depressing news, or seeing that their friends are doing things that they are not." To track patients' symptoms, MacMillan uses several measures, including the self-rated BDI-II and the clinician-rated HRS-D.

MacMillan says that people tend to stick to the protocol pretty well because they see it as a last resort for them, which it often is. A lot of them have
experienced failure of seven or eight medications, she says, or may even have
landed in the hospital because of their condition. Some of them have tried
the other last resort, ECT, and lost "chunks of their lives," as MacMillan
describes the subjective experience of memory loss. Her practice sees patients
who have tried ECT and endured "intolerable side effects, more memory
loss than they can tolerate." That said, she added that ECT gets a bad reputation because of its portrayal in movies as nonconsensual, but some people
with depression do volitionally choose it—an indication of just how bad the
depression can be. It's unfortunate, to put it mildly, that people have to clear
so many obstacles and even risk damage to their memories rather than have
access to interventions that seem to be so effective.

Before patients reach the point where they can access these interventions
because of the "failures" of other forms of treatment, many of them will
have encountered cognitive-behavioral therapy. In a way, CBT is like neurofeedback minus the feedback. The patient is the protagonist, the thoughts
and mood are the foes, and the patient's job is to replace them with thoughts
and moods that are friends. To do this, our protagonist must be what most
humans are born to be: a scientist.

Not in the sense of "earn a degree in chemistry" but in the sense of doing what our brains do best (at least, when they are at their best): taking in
information, analyzing it, drawing conclusions, and then taking in more information against the backdrop of those conclusions. If the new information
warrants a new conclusion, we adjust accordingly. CBT gives people a framework to follow so that they can take that last step and pry themselves out of
thinking and behavioral ruts that cause them problems. Doing this involves a
reappraisal of earlier conclusions, examining them from a perspective outside
of your feelings about yourself. You can still ruminate, but CBT is intended to
channel that rumination into something more fruitful and positive.[44]

So in CBT, we use a tool in our heads to retool our heads, thinking our
way through and possibly out of behaviors. Or at least that's what we do if it
works. Which it does not do for everyone, and, as always, study findings are

mixed and designs flawed.[45] We're not all in a place where we can access our executive brain sufficiently to apply beautiful analysis and reappraisal. For that reason, a lot of studies focus on combining CBT with other approaches, such as ketamine or psilocybin therapy or neurofeedback. The good news within this jumble of choices is that we have them. We have more than a single way to attempt to tailor our brains before we give up.

One method might be an actual tool for tailoring: needles. I am an acupuncture skeptic, but the landscape of evidence for acupuncture in mood regulation has changed in recent years. As an example, a 2018 Cochrane review revised its conclusions from a previous review because the meta-analysis pointed to a minimal benefit for depression of acupuncture over sham control treatment.[46] The authors were careful to note the low quality of the evidence.

Among the evidence were the findings from eleven trials that included 775 people taking medication for depression. These studies compared the effects of adding acupuncture or not to the medication regime. When the findings were pooled and analyzed, the result suggested that acupuncture, when combined with medication, was "highly beneficial" in generating positive effects on depression symptoms. The addition of acupuncture to psychotherapy in two trials with 497 participants did not show this apparent benefit.

In a study completed after that review was published, acupuncture did yield benefit in symptoms of depression. The authors allocated participants randomly to medication only or medication plus acupuncture and found that at four, eight, and twelve weeks, the acupuncture-added group had improved scores on the HRS-D.[47] The researchers also measured chemicals in the urine of participants, looking for changes related to breakdown of molecules involved in mood—including glutamate—and found that they were different in the acupuncture versus nonacupuncture group. The relevance of the differences is not clear. As with most interventions that involve an aspect of increased positive attention from other people and hints of virtuous self-care, a placebo effect is possible, even for objective measures such as urine metabolites.

MIXED METHODS MITIGATION

In another mix of methods, cognitive therapy has been combined with mindfulness practices as a way to regulate mood and treat depression.[48] In these approaches, the patient is again the protagonist. The intermediary can be a person guiding the meditation, or it can be the patient themselves. Meditation has shown benefit for depression symptoms, forcing as it does the engagement of the brain's executive network with mindful attention to making an intrusive DMN shut up.

Add this deliberate engagement to the practice of cognitive reappraisal, which, as you'll recall, involves a reconsideration of conclusions about past experiences and thus a reconsideration of one's personal role in those experiences. The protagonist also should be recalibrating conclusions, while, the hope is, dampening the negative emotions that paint them.[49] The other hope is that with practice, the ability to click away from the DMN's persistent negative messaging becomes something that can be done on cue, similar to the putative effects of neurofeedback.

Except in this case, the cue is awareness of what the DMN is doing, which involves paying attention to the unconscious mind-wandering that can take you into unwanted territory. Attention paid to the network's less positively adaptive tendencies, coupled with a practiced ability to dial it down and shift into the present moment, signals a flexible brain that's in control. In fact, research suggests that depression is associated with difficulty in achieving this toggling.

Mindfulness practices come in different flavors, but the established standard around which they vary was developed by Jon Kabat-Zinn in the 1970s and remains in use today. It involves a mix of group sessions and less than an hour per day spent doing individual work, all culminating in a group retreat. Variations might involve practicing gratitude or compassion.

The practice of compassion can certainly be turned inward, to yourself. We all need self-compassion, and we have to make room for it as a conscious part of our mood cognition practices. Evolution shaped our unconscious default mode to examine social experiences through the lens of what we did wrong and how we can avoid doing it next time. That's how we stumble into maladaptive rumination, and honing our mood cognition tools can be how we dig ourselves out.

Revisiting our social memories offers opportunities to practice this compassion for both ourselves and others. Almost all our memories are social in some way, involving other people or wondering what they would think if they were witnessing them. Say that the rumination serves up a particularly horrible experience from your childhood of having been bullied. You wonder if something personal and terrible about you attracted this negative attention and abuse. You blame the people who did it, too, and maybe even hate them. But remember that the anger, lashing out, and violence of bullying are often rooted in stress, anxiety, depression, and being abused, too.

If you can feel compassion for a child who tried to hurt another child as an outlet for their own anxiety, stress, and poor mood, you can gain two gifts. You can understand that person as more than a one-dimensional bully, which means you're experiencing empathy. And you can understand that their behavior toward you was not about you personally. You just happened to be there, and no personal failing on your part meant you "deserved" that treatment.

In this way, rumination, when consciously guided toward compassion, can be adaptive, for you and for the people in your life. To speak personally, in the process of writing this book I have tried these practices myself. The last few years have presented challenges in applying conscious compassion, requiring a lot of heavy lifting from my empathy tools. But those tools have helped me temper my anger and repulsion at reckless and selfish behaviors and think more clearly about the factors that drive them. To target a physical disease, we must first understand what's happening in our cells to cause it. Then we develop therapies that disrupt those damaging processes or repair the harms they cause. The same applies at a social scale: we need to understand what causes damaging behaviors before we can address those causes and reduce or repair the damage . . . or effectively defend against them.

Addressing causes often should mean social consequences for the perpetrators of these behaviors. Knowing why people act as they do doesn't confer blanket, unasked-for forgiveness on them. But adopting an empathic understanding protects me from the most terrible of reactions: depression resulting from an inability to fathom what motivates human cruelty. It's no mystery to me. I know it and understand it, and applying empathy across this gap in human behaviors means protection against being cruel myself. I also don't need to struggle with trying to forgive people who have not asked for it, which they usually should do as a first step in mitigating the harms they have caused. And no rule says you have to grant it.

Another challenge of recent years has been, of course, the coronavirus pandemic, which caused many of us to become socially isolated. As a return to in-person interaction became likelier, we began to experience social anxiety, thanks to our unused social cognition. Surveys suggested that quite a few of us tangled with mood problems, either new ones or the worsening of existing ones. We were an assortment of isolated DMNs, grappling with despair, and rusty on social cognition.

That's a recipe for mood disruption and anxiety. Loneliness, concerns about social ineptitude, and a shrunken social network all show associations with how the DMN operates in depression.[50] And they illustrate how interconnected our social, memory and attention, mood, and stress cognitions are. An intensely self-focused DMN compromised by depression can't play its role in social cognition. An overloaded executive and attention network can't attend to social cues properly, and activities that we used to perform on autopilot require manual engagement.

In our new normal, as we try to right ourselves and reposition our social roles, we need to remember what our different cognitions—our internal family network—have gone through. As millennials I know say, "It's a lot."

CHECKLIST

✓ Attend to your social cognition.

When addressing our mood, social cognition is probably the first tool we need to break out, particularly our empathy and its role in underpinning compassion for ourselves and others. We all make mistakes and feel awkward, even when we aren't dealing with the aftermath of a pandemic peak and paroxysmal political cruelty. In this chapter and those preceding it, you've been introduced to or reacquainted with a host of tools to consider adding to your personal brain-tailoring kit. Empathy and compassion may be the most important, and they are the only ones that I'd recommend most individuals keep always at the ready.

✓ Be not idle.

To show how little things change in human time frames, I take us back to Robert Burton, author of *The Anatomy of Melancholy*. He had a message buried in that tome that captures the human condition and how we can encourage its healthiest state: "Be not solitary, be not idle." He was not wrong. We found during the pandemic that taking up non-idle practices like baking or knitting or coordinating vaccine appointments for others kept us imbued with a sense of purpose and a lack of aimlessness.

These occupational therapies occupied our hands and minds. When we shared them with other people—feeding them the bread we made, joining knitting groups to share tips and tricks, helping others access preventive health care—we knew we were doing something positive. It felt positive because the effects were social. And it may also have diminished, at least a little, the chance of our being overwhelmed by a flood of anxiety and its companion, depression.[51]

✓ Let your freak flag fly.

As a bonus, these practices, including finding clever ways to schedule vaccine appointments for people, involved creativity. Our need for busy hands as a conduit to connecting with others also brought out our human inventiveness and desire to create. In the aftermath of the miseries and horrors of recent years, as the world struggles back to its feet, we will probably see a burst of human expression in the form of art in all media, new ways of doing old things, and scientific and technological advances.

From our despair, as we seek purpose—and perhaps occasionally microdose away our tendency to engage in obsessive self-examination—we will look outward and use the capacity that distinguishes us from all other animals: creative cognition. It has allowed us to develop the rich culture and technology that are the trademarks of our species. Interestingly enough for our interesting times, a personality trait that is closely associated with creativity is openness to experience. You'll read more about that in the next chapter.

CHAPTER 9

CREATIVE COGNITION: UNLOCKING INNOVATION

EVEN THOUGH, BY comparison to other animals, humans are enormously and uniquely gifted with creativity, a lot of us seem to think we aren't creative enough. Perhaps our outer lives don't reflect our inner desires, or we don't have time to fully explore our "creative side"—which in fact is not a side but several brain networks working together.

In this chapter, you'll see how the implements we turn to for creativity affect these important networks. And you'll learn how building the skill of engaging them all at once may be the key that unlocks that "side" of you. It is a skill you can practice, but like most of the things you've read about in this book, enacting it takes time. There has yet to be an intervention developed that buys us time, per se, but skillful work with neuroergonomic tools might carve out a few minutes for you here and there. If you consciously take those minutes to reach for creativity triggers, then you're doing it: you're working out that creativity muscle.

The study of creativity hasn't gotten the level of attention from our collective salience and executive networks that our other cognitions have. But what researchers have uncovered about creativity and the processes that might lie at its deepest roots fits with the symbolic language we use about creative output: edgy, at a tipping point, something resting at an angle of repose that a mere nudge can trigger into an avalanche. As you will see, when we experience the avalanche, we may be the least focused on ourselves that we can be, and we have quite a few accessible (and some illegal) ways to trigger the slide.

As you read about the various ways humans have tried to expand their creativity, you can develop a creative cognition. That's a way to consciously choose the tools that might bring you flashes of awe and inspiration, based on a current understanding of what creates creativity.

Reaching for creativity can feel risky, and methods to maximize it can *be* truly risky, to health or freedom. Some options are completely legal and low risk but possibly also low effect, while others might take you farther out than you want to go unless you exercise considerable care.

Whatever tools you choose, please don't go on a "psychonaut" journey by abusing "psychedelic fauna." Those frogs, fish, ants, and other animals* don't exist for you to capture and abuse just to boost your creativity.[1] Humans who have incorporated these animals into their local cultural practices are working as natural organisms within their natural environment. San Franciscans who capture Amazonian giant monkey frogs (*Phyllomedusa bicolor*) just so they can poke at them to induce a stress-triggered release of toxins to use in a euphoria-inducing barfing "detox" are not. The former commonly use such substances for their destructive effects on pathogens. Silicon Valley types, on the other hand, are doing it for, as one practitioner put it to the *New York Times*, "bohemian masochism."[2]

Before we reach for creativity, perhaps it would be good to know exactly what it is we're trying to grasp. Neuroscientists home in on three key features when they consider creativity: originality, surprise or rarity, and relevance or appropriateness.[3] You may notice that those three factors happen to be the same qualities in events and in our environments that arrest our attention,

---

* The list is quite long and includes the clownfish, the convict surgeonfish, sea sponges, and a species of harvester ant. Apparently, we humans really will try anything. The commonality among the animals tends to be that they use the psychoactive substance as a defensive toxin.

help us learn, and aid us in sorting out what is salient and what isn't. In other words, creativity is what our brains crave from the external world. The element in our environment that serves it up most often is other people.

Creativity then, is a social interaction, so it's no wonder that as a social species, we love it when we can generate it ourselves. We can both express and experience it. It's the cognitive quality that distinguishes us as the species that can communicate across time and space with others we have never met. That also means creativity doesn't have to be splashy or multicolored or outré. Just novel, leading us to learn and refit our frameworks. In addition, creativity is situational. If relevance or appropriateness is a criterion, then by definition, environment—include society and culture—defines what creativity is.

Broadly speaking, creativity as a process in the brain can be classified in two ways.[4] One is divergent thinking, meaning your ideas veer off into nonlinear territory and make unobvious, unexpected connections with other concepts. That's a freewheeling form of creativity, somewhat like brain-storming. The other kind of creative thinking is convergence. In evolution, convergence occurs when unrelated species develop similar features that are used for the same purpose, such as the very different types of wings on bats and butterflies, all of which are used for flight. In terms of thought, convergent thinking means arriving at the same solution by many different paths. Divergence requires releasing your sense of self to an extent, as seems to happen to people using psychedelics. Convergence is more analytical and calls for attention and executive function for their roles in problem-solving.

Regardless of which type of creativity is in play, the common theme that binds them is flexible thinking. You can't diverge if you're in a rut. You can't see the many paths to a solution if you're fixated only on the one before you. Being flexible means having the trait of openness to experience.[5] It is also the characteristic that can make or break social progress and acceptance of change, and that can open the way to empathy and perspective-taking.

One drug that's often associated with creativity and the feeling of open-ness is psilocybin. Robin Carhart-Harris's group at Imperial College London has done a ton of psilocybin research and found reduced authoritarian tendencies in people who took it,[6] making them more liberal. If "liberal" is the opposite of "conservative," which definitionally seeks to preserve the status quo, then being liberal is being open. The results of one study have hinted as much, showing that people with what the authors call an "open-minded thinking style about evidence" also tend more often to identify with a liberal political party.[7]

One of our most consistent manifestations of creativity, across cultures and time, is storytelling, the combinatorial art of creativity, originality, analytical and attentional processes, and relevance to common purposes. It is also the act that ties together all our cognitions: global, social, memory and attention, stress and anxiety, mood, and creativity. The convergence lies in the blending of these cognitions into a creative expression that is crucial to our species. Our stories come to us visually, verbally, musically, and through movement, taste, and touch. We can tell them in countless ways. The common factor is the sharing. Without the audience, storytelling has no purpose. And if we lose our stories, as the cautionary tale of *Fahrenheit 451* communicates to us, we lose our human touch, literally and figuratively. Empathy dies on the vine, and our humanity goes up in smoke.

IMAGINATION STATION

The novel *Fahrenheit 451* sprang from the vision of that master of imagination Ray Bradbury. The dystopian tale of a future America where books are banned* reflected the real threat of McCarthy-era censorship and the human disconnection that followed the death of storytelling. The result of the attempt to block stories was a homogenized, ignorant, and insouciantly oblivious society guided and monitored in the most depersonalized, dehumanized way—through technology. The threat of censorship at the time was, of course, real. Bradbury was not the only artist to create a parable of the hysteria related to the era's witch hunts for alleged communists: Arthur Miller's *The Crucible* is another well-known example. Both men, though, wrote of worlds that were not their own, one set hundreds of years in the past, and the other set in a future that had no obvious prelude.

To achieve their feats of imagination, they had to engage three key processes:[8] simulation (creating representations of how people might behave

---

* 451°F is the temperature at which paper ignites.

in specific scenarios, taken from experiences encoded in memory), mental time travel (back to the late seventeenth century or forward to 2049), and perspective-taking (putting themselves into the inner lives of their characters, protagonists and antagonists alike). Each of these is specifically a human capacity, as far as we know, and together they form the trifecta that fuels imagination and creativity. In the brain, they are hypothesized to arise from two related systems.[9] One is our mind's eye, the agent that allows for free-floating, off-task ruminating (not necessarily the negative kind), during which we time travel and freely envision without conscious guidance. The other is our mind's mind, where the analytical bit kicks in and allows us to consciously contextualize our daydreamed imaginings, giving them coherence and shape.

When we express the result in whatever medium we choose—the spoken or written word, pictures, motion—we are being creative. It has been proposed that the mind's eye consists of a subset of the DMN that acts together with the medial temporal lobe, which serves up visions. The mind's mind is suggested to consist of another subset of the DMN,[10] connecting with the dorsal medial prefrontal cortex, where much of our social cognition resides.

The authors who proposed this construct in 2021 not surprisingly gave examples related to the COVID-19 pandemic, using images and words in a cartoon. The cartoon depicts a person with two thought bubbles over their head, one representing the mind's eye, and the other representing the mind's mind. The mind's eye, which is described as producing output that is "contextual, detailed, concrete, specific," is thinking, "I can picture eating ice cream down by the beach at sunset with my friends." Meanwhile, the mind's mind, generalizing this daydreaming into something social and more broadly applicable (and described as abstract, general), is thinking, "I wonder if my friends are as lonely as I am during COVID-19." This example illustrates how imagination both relies on and fuels our capacities for time travel, perspective-taking, and simulation.

Leveraging imagination can be risky, whether we do it unconsciously or deliberately, whether powered by our own minds or with an assist from a nip of a psychedelic. The reason reaching for creativity feels risky is that it opens us up to uncertainty, to realizing or discovering or learning something that makes us uncomfortable. That's where openness comes in handy. If we're receptive to originality, there's less leveling up involved when we confront creativity. Having a temperament that is open to novel experiences can, well, temper the shock of the new. Some people react to "shock art" such as *Immersion*, by Andres Serrano,* with a shrug or maybe a bit of disgust at the portrayal of bodily fluid, whereas others are affronted and outraged at the treatment of the crucifix and what it symbolizes, viewing it as blasphemous. Openness can mean you've set a higher bar for such strong reactions. But it also can mean being more demanding than the average human about what qualifies as creativity. As I noted above, creativity is situational.

Our brains, when operationally at their best, may spend much of their time at this threshold, edging near some cliff of creative freefall but usually balanced in equilibrium unless an event, cue, or trigger tilts us over. Some researchers hypothesize that when we maintain this stability a little too well, we "overfit" our thinking, placing our conscious thoughts into a constrained safe zone that's comfortable and familiar. In this way, we go about our day, our autopilot largely in charge, our attention often unpracticed and perhaps a tad dulled.

AVALANCHE

Dreams, which happen only in our heads, present a safe way to break up this monotony. In what researchers term "noise injection,"[11] our dreams may give us the push we need to set off an avalanche of impossible, chaotic stories that, if real, could lead to our harm. We fly, we fall without dying, we take risks that would be life threatening in the real world. We may be consolidating reality in our dreams in some way. But around that core, we diverge like nobody's business, trying out all kinds of illogical alternatives, solutions, and scenarios. These ideas are untested and new. But they are interesting for our purposes because of their reflection of another new set of hypotheses about the brain. In dreams, we tilt past the autopilot that orders our day and into the chaos of a distorted reality.

Perhaps you've heard of the angle of repose, the maximum slope that a pile of loose material can maintain without destabilizing and falling. Think

---

* Also known as *Piss Christ*, it is a fireball-tinted photograph of a plastic crucifix immersed in a container of the artist's urine.

of making a pyramid of sand. The steepest its sides can be before the sand collapses is the angle of repose. The structure is in a state of order, but with the teeniest nudge or the slightest addition of more sand, the whole thing would tumble down in a state of chaos.

In the brain, a similarly precarious stability may exist that, if disturbed, could set off an avalanche of motion. In this case, the disorder is in the form of chaotic activity bursts that diverge from the usual rhythmic oscillations. The brain can be in a state somewhere away from this critical point, somewhere quite near it in a precarious balance, or past the brink and experiencing the avalanche. The shift from "precarious but stable"—which may be the optimal state for taking in and sorting through information without being overwhelmed[12]—to "avalanche" is called a phase transition. This transition and the resulting turmoil may awaken creativity.

The process has been compared to what happens when a lot of tiny earthquakes combine to set off a larger one. First, tension in some small region of the Earth causes it to surpass its angle of repose and shift, setting off a minor quake. That tremor in turn pushes adjacent areas past their angles of repose, and they also shift. Collectively, this chain reaction can activate an effect on a large fault, causing it to exceed its angle of repose and triggering a huge quake.

In the brain, a cascade could start with a single neuron and ultimately amplify to involve a huge swath of territory and millions of neurons, causing the avalanche. Up to that point, perhaps there were rumbles as the cascade reached the edge of the tipping point but never quite shifted phase. According to this hypothesis, that is the optimal state: stably edging up to the shift without going over.[13] Maintaining the brain just short of this cliff keeps us consciously aware of and alert to information as needed without being overwhelmed by all the inputs; we are balancing incoming signals of many varieties and strengths against our internal models that make sense of everything.[14]

When we tip over, many things can happen. Sometimes the brain finds itself in a maladaptive state that becomes dominant, as may be the case in conditions such as schizophrenia. But if we make mere forays over the edge and then hustle back to the other side, creativity can follow. And there are quite a few things we can do to nudge ourselves over without going into permanent free fall.*

Not all experts embrace this concept,[15] which I've only sketched out here and which remains in the early stages of development. But it seems apt for our chapter on creativity. Studies suggest that psychedelics such as psilocybin and LSD are linked with brain-imaging markers of being at the tipping point[16]—or passing it. These findings indicate that when we're perched at the brink, neurologically speaking, our consciousness is at peak normal. We experience dissociations or changes in consciousness if we either back away from the tipping point or dive over.

As a species, we may have entered our creative period about fifty thousand years ago, when the modern version of us became established. Our ancestors began to produce tangible manifestations of this uniquely human power, including pigments that they used to create art and symbols that clearly told stories we can still decode thousands of millennia later.[17] As we built up this skill, we also developed brains that were optimized in key networks associated with the skill. These networks included regions related to attention, planning, imagining, sensing the self, and long-term memory, among others—all of which feed our creative cognition.

In addition to the prefrontal cortex, two regions in particular became associated with the modern human brain: a comparatively large cerebellum, which has many functions, including contributing to our social cognition and working memory, and a bulging parietal area. The structure causing this bulge is the precuneus, a major node in the DMN.[18] It lies in the posterior area of the parietal lobe, which, along with other DMN nodes, is associated in neuroimaging studies with the generation of divergent thoughts.[19] The DMN is our guide when we time travel and review our autobiography, both crucial elements of creativity and imagination—and possibly uniquely human.

But we also know that other aspects of creativity demand our attention—that is, they require us to engage our attention and our executive function networks. So it makes sense that studies implicate not only the DMN but

---

* Neuroscientists can use a readily available universal mathematical function to assess these hypothesized behaviors and don't much care for it when people use their specific terms for this construct, such as "criticality," to represent something more symbolic.

also these other networks in creative thought processes.[20] The DMN serves up the divergence, perhaps, taking care of the "novel" part of creativity. The executive and attention circuits sort out what's reasonable, handling the "appropriate and relevant" aspect.

When the brain engages in that divergent thinking, imaging studies suggest that the DMN is in close communication with the hippocampus, probably to summon episodic memories to review.[21] Interestingly, in light of all the studies that seem to confirm that the DMN and the executive control networks anticorrelate—that one goes quiet when the other is loud—creativity studies suggest the opposite: during imaginative thinking, both networks are active, along with the salience network. In other words, when you're being creative, you're revving up three networks at once.[22]

It's possible that specific regions of each network are most engaged, a signature of creativity that one group has reported.[23] Indeed, in an imaging study of 163 people completing a common divergent-thinking task, researchers found that all three networks showed activity in their frontal and parietal (there's that lobe again) nodes.[24] Investigators also found similar coactivation of the DMN and the executive network in poets and visual artists while they were engaged in their creative processes.[25] These authors concluded that we have a "whole-brain network" linked to being highly creative. It uses selected hubs from these three networks—which usually operate as "frenemies" at best.

CAPTURING CREATIVITY

Recall the chapters on global cognition. Perhaps you've now realized that the tests for this capacity don't usually involve a creativity component. That's kind of odd, given that creativity is so crucial to our advances as a species (and to getting us out of at least some of the problems we get ourselves and the planet into). It's also odd because creativity is associated with better outcomes related to education and to work and in the arts and sciences, which has often been the putative rationale for having people take IQ tests.[26] The closest thing we saw to an effort to build creativity were the fun problem-solving tasks that Radivoy Kvashchev gave to the Belgrade high school students to work on together.

But to study creative thinking, some kind of tests are needed. Unlike so much research in neuroscience, creativity studies show a little more consistency in the measures they use. One of the most common instruments used for divergent thinking is the Torrance Test of Creative Thinking. It relies on verbal or "figural" cues in which a basic construct is presented and the test taker is instructed to do what they'd like with the stimulus, within specified limits that depend on the test activity. For example, in a picture-construction activity, the test taker might be given a shape that has no specific purpose, such as a pear, and be instructed to make a picture that incorporates the shape. A trained scorer grades the output on its originality, abstractness, and amount of elaboration, and compares it against a checklist of creative strengths.

A common test used to assess convergent thinking is the RAT, or Remote Associates Test. As the name implies, it presents things that are associated, but in a remote way, not obviously (no "person, man, woman" here). The test taker is given three words and has to figure out what they have in common. An example is "bald/screech/emblem" (see footnote for answer).*

In a more complex version of the test, the test taker has to not only identify the commonality among the three words but also find the association that creates a compound word out of each.[27] Consider the following examples, listed here in order of difficulty from "easy" to "most difficult" (answers are in the footnote):†

✓ cadet/capsule/ship
✓ wheel/hand/shopping
✓ cross/rain/tie
✓ fence/card/master

As you can tell, the expectation is that the person taking the test is someone who grew up in the United States speaking English, as the answers don't translate to other languages, cultures, and backgrounds.

---

* The commonality among them is "eagle."

† Answers: Cadet/capsule/ship = space. Wheel/hand/shopping = cart. Cross/rain/tie = bow. Fence/card/master = post. You can find a version of the test here: www.remote-associates -test.com/.

Because these tests are fun, I'm going to tell you about another one, called a manipulative insight problem. It falls into the category of convergent creativity. If someone gives you a candle, a matchbox, and some tacks and tells you to fix the candle to the wall so that it can burn safely without dripping any wax onto the floor—what would you do? This test addresses the mental ruts we can get into when thinking about objects that have a familiar purpose, or functional fixedness.[28] If we can break out of seeing such items as serving only their usual purpose, we can probably solve this problem.*

Yet another test, known as the alternative uses task, gets at divergent thinking, and it's the most brainstorm-like of the bunch. The test taker is presented with the task of thinking of as many uses as they can of a common item, such as a paper clip, shoe, or spoon.[29] Scores are based on how many ideas the test taker can come up with in the time limit, with extra points for multiple categories of use and for originality, especially if the suggested use is unique. This task is divergent because it has many different possible outcomes, as opposed to a convergent task, which has one possible outcome arrived at by many potential paths, like the candle test or the RAT.[30]

Creativity has been linked to a host of favorable consequences that many of us might be interested in achieving, including enhanced sexual desirability and better physical health.[31] That matters because creative cognition allows us to see when an idea is novel and relevant and to embrace the change it promises—in other words, to be open to new experience. People who score high on "openness" also report having good imaginations and, beyond the subjectivity of self-report, score in the high ranges on tests of creativity.[32]

A well-functioning DMN has been linked to openness. Some research suggests that whatever openness we have tends to decline with age, in keeping with what seems to happen to the DMN as well—and in line with the observation that we become more conservative with age. Perhaps if we consciously engage our creative cognition, we can maintain our openness—and our creativity—as we get older.

In my sojourn through the research on what drives or enhances or harms creativity, the last thing I expected to come across was tuberculosis

---

* In this case, you can tack the matchbox to the wall, and fix the candle in place with a little wax that you melt off using a lit match.

as a candidate. I have to give Arthur C. Jacobson, the fellow who proposed this association in 1908,[33] props for achieving two of the key features of creativity: surprise and novelty. Appropriateness, not so much. He made the argument that because tuberculosis plagued many famous writers, from the Brontë sisters to Jean-Jacques Rousseau, the disease must have been the source of their creative output. His hypothesis—accompanied by a great deal of armchair diagnosis of tuberculosis itself—is obviously a failed one.

He argued "in an Emersonian spirit" that the early loss of certain "geniuses" was a welcome trade-off for the creative fruits they bestowed on the world. Charlotte Brontë, who did not in fact die of tuberculosis, as Jacobson asserted, but probably of hyperemesis gravidarum during early pregnancy, would likely have disagreed. Watching her two adult sisters die of the disease, Emily in particular angrily fighting it every step of the way, left Charlotte traumatized.

As writers, Emily and Charlotte Brontë both captured with realistic immediacy the power of the natural world: a lightning-scarred tree, a rocky moor scoured by the wind. Or consider Emily's portrait of imagination in her poem "To Imagination," which she likened to "a bright unsullied sky / Warm with ten thousand mingled rays / Of suns that know no winter days." They communicated with immense skill a sense of awe and how nature and the supernatural humbled them while simultaneously filling them with gratitude. Tuberculosis had nothing to do with it.

THE AWE EFFECT

Their authentic communication of this awe may have contributed to their creative brilliance. They are best known for their novels and poems. But when they were young, these two sisters, along with a third sister, Anne, and their brother, Branwell, spent years creating entire worlds in the minutest detail. In tiny script, they wrote in tandem, inventing characters and stories to populate and fill out the history of these realms, including the fictional kingdoms of Angria and Gondal.* The worlds were as complete as anything

---

* The invention of these kingdoms was a breakaway rebellion on the part of Emily and Anne, who became separatists from the worlds established by their older siblings, Charlotte and Branwell.

Tolkien ever devised, with maps and fully wrought characters, some of them based on real people.

The Brontë family, largely confined to a lonely parsonage, was stalked by tragedy. But they were surrounded by an astonishingly beautiful patchwork of heath and stone, a moody and changeable setting for expression of the internal suns that knew no winter days. Perhaps the stark contrast between this natural beauty and the darkness that shadowed their lives fed their sense of awe, sparking the creativity that made them famous despite their short time on Earth.

I'm not talking about the Brontës just because I'm a fan.* I'm talking about their communication of awe to point out that if this family, living beleaguered and tried by loss after loss, could access awe, then many of us should be able to. And awe can well be one of the energy sources for the ten thousand mingled rays of the suns of imagination in each of us.

Imaging studies show that when people experience awe, the DMN becomes more loosely connected compared to its activity during a neutral or even a positive experience.[34] To tease this out, researchers showed study participants three types of videos: awe-eliciting, positive (funny animals), or neutral (basic, un-awe-inspiring landscapes). To measure the whole brain during "absorption," they had the viewers focus on being fully absorbed in what they were seeing. To measure the brain during a task, the "analytical condition," the participants were instructed to count the number of times the camera changed perspective. At the same time, the study volunteers also reported on their feelings of awe, emotion, and arousal (in the hyped-up sense, not sexually). Absorption, or being off task, not surprisingly was associated with more activation in DMN regions. But when the video was awe-inspiring, the activation was weaker than when participants were being absorbed in positive or neutral videos. The feeling of awe hushed the DMN a tad.

When the participants were on task during the awe-producing videos, they showed stronger activation in regions associated with attention compared to the effects of counting perspective changes during the neutral or positive videos. The findings suggested that the DMN remained active

---

* Although I am.

during awe, but less so than in everyday absorption, even as the salience network gained in strength during a feeling of awe.

The authors concluded that awe dampens our self-referential thought, in keeping with the feeling of being "humbled" that people express when they encounter something wondrous, such as a grand view of nature. The vastness of the moors that surrounded the Brontës' home certainly could have inspired such reverence. Indeed, vastness itself has been linked to reduced self-referencing.[35]

As we have seen, experiencing the natural world can have salutary effects in all kinds of ways. These studies hint at a role for nature in our creativity. With this information in hand, you've got a tool in your creative cognition toolkit that may lie just beyond a closed door or window, depending on where you live.

AWESOME, PSYCHEDELICS

Probably not waiting just outside your door is someone offering you psychedelics, although that also depends on where you live. These drugs may interact with facets of the creative muse. In a small study of sixteen people taking an LSD trip, their scores for trust and openness went up after their journey (although so did their blood pressure and heart rate). The participants also felt closer to other people, leading the authors to dub LSD an "empathogen."[36] In a 2021 double-blind, placebo-controlled study of sixty people split into drug (psilocybin) and placebo groups, the psychedelic seemed to enhance divergent creativity, strangely on a seven-day delay, but dampen the more deliberative, convergent kind of creative thinking. These researchers also found a "disintegration" of the DMN associated with the increase in divergent creativity.[37]

Tripping on psilocybin has been associated with higher scores on measures of "openness to experience," and the effect was especially long-lasting if the traveler had a mystical encounter during the session.[38] In fact, in one randomized, double-blind, placebo-controlled trial, people who reported a mystical experience while taking psilocybin were more likely to experience mood improvements that persisted weeks later, suggesting some dimming of the DMN's insistent "me-me-me" drumbeat in depression.[39]

In keeping with research showing that engaging our creativity triggers two or three networks that normally operate at odds with each other, findings with classic psychedelics show that the executive network and the DMN get into sync under the influence of these drugs as well. Perhaps this syncing contributes to the hallucinatory confusion about what's real and what's all in the brain.[40]

In an echo of the precariousness of the brain's rhythmic tipping point, the authors of one review say that their findings may point to "increased instability in psychedelic states."[41] Another group of authors found that alpha waves, which are supposed to keep attention directed toward important things and to filter out random noise, decrease in power under the influence of LSD.[42] No more blocking of unimportant information.

Before Impressionism existed, no one knew what it was. When it was introduced, it was novel. That made it creative. One of the things the Impressionists did was to "liquefy the solids and corrode the angles."[43] It's difficult to read this allusion to both state transition and altered angles without thinking about the brain's hypothesized tipping points and how they relate to creativity. Impressionists struck the world with their reshaping of visual reality; psychedelics similarly alter our sense of reality. Many of us, in the absence of a Brontë-like power to summon experiences and entire worlds from imagination alone, need something to nudge us over the creative edge.

There's more going on with these drugs than some hypothesized or symbolic tipping point. In the very real but invisible molecular world, as we saw in our chapter on mood cognition, ketamine and the classic psychedelics show a propensity to set the stage for neuronal excitation thanks to the effect of glutamate.[44] The excitement isn't universal: one study showed increased glutamate in the prefrontal cortex in people taking psilocybin, but a decrease in the hippocampus. That scenario suggests some effect on memory delivery and, these authors concluded, a stepping away from a focus on the self.[45]

Perhaps this surge, even if it occurs locally, sets off a cascade of reactions until it encompasses the whole brain, like a series of small earthquakes initiating a big one. Arguing against that hypothesis are findings that brain waves in the low frequencies show weakening under the influence of psychedelics, with a possible trend to more local communication.[46] The conditions that can tilt a brain past the tipping point, if such a state exists, are complex, and many pieces of this hypothesized puzzle remain to be uncovered.

No mind-altering substances other than the psychedelics achieve this mix of released ego, mystical experience, openness, and divergent thinking. You might think alcohol could unleash the ingenuity fairy, but randomized, placebo-controlled studies* don't find much effect. After one drink, study participants may get more creative about solving problems, but beyond that they just get impaired.[47] There may be veritas in vino or beer, but there isn't much invention.

YOUR MOTHER WAS RIGHT

Could I interest you in your fruits and vegetables, particularly their fiber? If you're skeptical, stay with me. In fifty-seven children and teens whose dietary intake was monitored for three days and who then took the Torrance Test of Creative Thinking, high sugar intake was negatively associated with creativity. But high fiber intake was positively associated with their performance on the test.[48] Bear in mind that the people in this study were children, who are not just small adults and who have their own developmental needs. What about adults?

In another study of 405 young adults who kept a diary of their intake of fruits, vegetables, sweets, and chips for thirteen days, fruit and vegetable intake was associated with greater curiosity and creativity.[49] This study was only observational, and the creativity and curiosity measures were self-reported, so other influences certainly may have been at work. Perhaps young adults with greater access to healthy foods have other advantages that help promote creativity, for example. The participants also reported having a better overall mood, which could be another factor. Still, the authors argued that the link between creativity and eating fruits and vegetables was "biologically plausible" because of the nutrients in these foods and their potential effects on plasticity, which is a big reach, and even a connection with the gut microbiota, which is an even bigger reach.

---

* There is nothing under the sun that someone hasn't done a study on.

A food-associated molecule is the amino acid tyrosine, which is what our bodies use to make dopamine. In a study of the effects of tyrosine on creativity test scores, one group found that compared to placebo, a dose of tyrosine was associated with better performance on a convergent thinking task (the RAT).[50] But the thirty-two people in the study showed no differences in divergent thinking scores between placebo and tyrosine administration. In any case, the authors were emboldened enough by their findings to conclude that "you are what you eat." Here's a twist: one reason they hypothesized that tyrosine might have some effect on creativity was because Steve Jobs, whom they called "arguably one of the most creative minds of our time," considered his diet based on apples and carrots to be "the foundation of his success." Both of these foods are high in tyrosine.

The same research group also took on the question of how physical exercise affects creativity. If you've been mentally arguing against every chapter in this book urging the benefits of physical activity for the target cognition, this is your moment. These authors found that exercise interfered with divergent thinking in their participants, forty-eight of them athletes and forty-eight of them nonathletes.[51] Nonathletes also experienced impairment in convergent thinking, although athletes did not.

But hold on. That study was published in 2013. A 2019 systematic review of this association found a little bit of there there. In the thirteen studies examined, although most of them suffered from methodologies that could introduce bias, the authors of the review still suggested some support for a benefit of exercise on creativity. OK, it was "weak to modest" support.[52]

They also wrote that the "lack of experimental work on creativity" as it relates to exercise science and health promotion "is staggering." They were surprised to find so few studies to include in their review, half of them published in the previous century, and most of them flawed in significant ways. Some of the studies didn't even cite how many people were included in the experimental and control groups.

Science really can be a hot mess.

For boosting creativity, you can probably skip neurofeedback and intracranial fritzes. A few studies examine both,[53] but the results aren't impressive—at least not yet. A study evaluating the effect of transcranial direct current stimulation on creativity in jazz experts found that the stimulation actually *hindered* improvisation in musicians with more experience.[54]

The practice of mindfulness is less risky for most people and, with the will in place, more accessible than transcranial stimulation or other brain-computer interface techniques. As one group reviewing published studies wrote, "There is solid evidence to show a generally beneficial and supportive relationship" between mindfulness and creativity.[55] The two may interact, with one supporting the other. Certainly, with its emphasis on nonjudgmental perception, mindfulness encourages openness and reduces the focus on the self, both of which are factors in creativity. A 2016 meta-analysis of studies examining the association found some relationship between the two—not a huge correlation, but not an irrelevant one.[56] The authors looked at eighty-nine potential associations in the data from twenty studies dated 1977 to 2015. In addition to finding a small but significant link between mindfulness practice and creativity, they discovered that the open-monitoring facet of mindfulness (avoiding immediate evaluation) carried far more weight than the awareness aspect in effects on creativity. There's that openness again.

Another group looking at seventy-seven college-student participants found that deliberately allowing the mind to wander* was positively associated with creative performance, but spontaneous mind-wandering was not.[57] The difference between the two, we could infer, is that deliberately activating the process involves our executive and attention networks, whereas the spontaneous version relies on the DMN. The researchers also found that nonreactivity and awareness were tied to creativity.

A study of creative writers seemed to confirm that this deliberate approach, rather than just unconscious roaming, was the source of their creativity.[58] This distinction tells us that we can indeed employ creative cognition, taking intentional action to boost our creativity rather than passively leaving our inventive processes in the hands of fate.

---

* Not precisely mindfulness per se, but related to it.

Does it really matter how creative we are? In addition to the associations that have been found between creativity and some measures of success in life, pursuing your creative skills could mean something positive for people beyond yourself. We create outside of ourselves when we manifest our ideas in the physical world. Steve Jobs didn't build and sustain Apple on his own. He was backed by a legion of creative brains all working together toward a common purpose. As a result, I am using one of their products to write this book.

As we've observed, creative impulses make it possible for us to communicate across thousands of years, which we did in the opening story of this book, about how a clinician in ancient Egypt treated head wounds. And creativity is likely a key to our success as a species. Research shows that it can be tied to survival, even among nonhuman animals such as some bird species.[59]

Our special ability to pass the fruits of our creative cognition down to the humans that follow in our steps allows knowledge to accumulate. And it has accumulated, generation over generation. If every new generation had to figure things out all over again, humans would stagnate. Without the products of communication that we've developed as a social, creative species, we'd probably be living much as archaic humans did three hundred thousand years ago.

CHECKLIST

✓ Awe
✓ Fruits and vegetables
✓ Nature
✓ Mindfulness
✓ Not alcohol
✓ Physical exercise

CHAPTER 10

A FREAKY FUTURE: FAR OUT BUT NOT FAR AWAY

Sooner than you may realize, things are going to get downright far out for our brains. If we don't have our DNA snipped with CRISPR before birth, we may find ourselves modifying our minds in other futuristic ways. All signs suggest greatly refined brain tailoring as a feature of our futures.

The open question is whether things will go the way of *Flowers for Algernon*, in which the protagonist experiences the best brain he's ever had, only to remit and return to his original condition or worse. With headlines like the one that appeared in the *New York Times* in 2019—"One Day There May Be a Drug to Turbocharge the Brain. Who Should Get It?"—these issues seem to exist nearer in the future than we may have imagined. Just to tone things down a notch, the counterpoint to the first sentence in the *Times* headline is "People have been working with this protein for decades and still don't have such a drug." The answer to the question the headline poses is "Anyone who'd like it."

No brain intervention comes without caveats. Sure, a turbocharged brain sounds appealing to some, but isn't that what many people suffer already?

Whether you describe the feeling as bees unleashed from a hive or utter mental chaos, the inputs we sustain all day and every day from an increasingly stressful, sometimes aimlessly busy world are turbocharging enough for me, thank you very much.

That said, if we're lucky, we get older. And an aging brain shows changes in the DMN and in its metabolism, changes that we might juuuust be on the horizon of warding off. The hottest thing going is brain stimulation—deep, micro, implanted. I am starting to think that in a few years, we may have brain-stimulation centers on as many street corners as we currently have Starbucks. Indeed, right now I could easily purchase my very own portable cranial electrotherapy device straight from the manufacturer for "only" $299. (Limited time offer for summer, says the manufacturer's website; the usual price is $379!) It comes with electrodes! You'll be relieved to know that I would have to get a prescription from a licensed medical doctor first. Thankfully, the manufacturer has one "Dr. Harold Stecker" to hand for such purposes should I be unable to find my own.

I think memory research is where we will see neuroscience matter most in the aging context. Research is going gangbusters, not only on memory in connection with Alzheimer's but also with the memory loss that accompanies aging. If the FDA were to approve a successful drug, I would take that pill in an instant to have a memory as sharp as I think it was when I was younger (I can't remember). The good recent news is that despite encroaching age and its effects, our brains may continue to produce fresh, shiny, new neurons (although the debate around this remains fierce),[1] which likely will serve as leverage for more antiaging interventions.

When we peek into our mental crystal balls, we are engaging in what you can call "episodic future thinking,"[2] in which we imagine unreal but detailed scenarios that have yet to happen. With our powers of future cognition, we base these predictive scenarios on the information we have at hand, and for the material in this chapter, that's the best we can do. Some people (cough, Elon Musk, cough) are extremely optimistic that before too long, we will be able to slip implants into our brains that boost recall, help us "inject" memories, perhaps limit aging-related effects, and even connect us virtually with other brains.

We can already perform some of these feats without the implants, but the constructs are clunky and awkward. Compared with what imagineers like Elon Musk envision, the hardware in use today is like a 1980s cell phone matched up against a 2021 smartphone.

The smartphone is already a machine we use for human-machine interfaces, so welcome to the future.

ROBOTIC MIND MELD

In 1988, in one of the earliest demonstrations of using brain waves to move something, the volunteer making the waves entered a state of relaxation. When the person was sufficiently relaxed for the brain to produce alpha waves, a computer processing the signals detected the waves and essentially "flipped a switch" on a robot it was connected to. This switch activated the robot's preprogrammed behavior.

Yes, a person relaxing with some electrodes glued to their head used their mind to switch on a robot through a computer intermediate. The robot in this case was a commercially developed toy kit that the researchers had purchased in Japan.

Obviously this kind of control is not very refined, relying on crude detection of waves with a signal strong enough to register through the human's admittedly thick skull. A goal of BCI research ever since has been gaining more refined control, and the field started to hit its stride as the twenty-first century approached.

Today, computer interfaces can detect neurons firing in several areas at once across the cerebral cortex, known as neural ensembles, and send this electrical information to a computer to be decoded and transmitted to, for example, a robotic arm. The arm apparatus, in turn, communicates information about touch and position back to the brain. The feedback is crucial for utility of devices that can reach and grasp, making fine-tuned adjustments as needed. Research has extended to using more than one brain (in rodents) working in concert to achieve a form of this "telekinesis"—which does freak people out a little bit.

Elon Musk also freaked people out a little bit in September 2020 when his company Neuralink produced a pig that he said had 1,024 robotically embedded electrodes distributed across thirty-two polymer threads in its brain. Musk had previously reported installing ninety-six threads, each bearing thirty-two electrodes for a grand total of 3,072, in the brain of a rat.[3] Pigs probably make for better performers on a world stage.*

Each electrode implanted in the pig's brain recorded the spike of a single neuron and fed the information to a computer. One of his demo pigs, Gertrude, had lived with her implant for a couple of months. As she snuffled around, the electrodes associated with neurons related to her smelling activity fired, and were detected wirelessly. In theory, then, signals from Gertrude's individual neurons could be recorded and used for . . . whatever purposes Elon Musk imagines and the "chaotic internal culture" of his company allows.[4]

Someone in the audience at the demo asked Musk if the technology could eventually be used to record memories and replay them. That application seems fraught with pitfalls, including the possibility of altering the memories (more on that below). One acknowledged pioneer in BCIs, Miguel Nicolelis of Duke University, professed on Twitter to be unimpressed with the Gertrude demo, pointing out that he and his team had executed a similar feat as far back as 2007.

The real advance that Gertrude displayed may have been the robotic procedure for implanting the thousand-plus wee electrodes to perform what were essentially micro-EEGs just outside each neuron, measuring local fields instead of picking up broader surges and ebbs from outside the skull.

THE DEPARTMENT OF DEFENSE IS STILL INTEREST ED IN YOUR BRAIN

One of the funders of Elon Musk's Neuralink/Gertrude the pig project is DARPA, the Defense Advanced Research Projects Agency, which you met in the chapter on mood cognition. You will definitely experience shades of *Minority Report* from some of the work that DARPA does. But if there's an arrow pointing at the future of the human brain, DARPA is holding the bow.

---

* An earlier attempt by this company to implant more than ten thousand electrodes into a sheep brain failed completely.

In 2017, Justin Sanchez, then the director of DARPA's Biological Technologies Office, gave a talk about the agency's projects.* Pony-tailed and wearing a suit, he reviewed the various technologies in development to "sense signals over time" from people's brains and, more hauntingly, to "write signals into the brain." The proffered reason for DARPA's interest in such projects is that military personnel sustain traumatic brain injuries and paralysis injuries at relatively high rates. Obviously, technology that could create a signaling pathway from the brain to the leg or from one part of the brain to the hippocampus opens a door to restoring movement or memory.

Predating Musk's Neuralink performance a few years later, DARPA funded researchers at UCSF to develop the robot that "sewed" those 1,024 electrodes on polymer threads into the brain. DARPA's mantra is to "create opportunity by reducing technical risk," and their "sewing-machine robot" is just one of many examples of the tech they're funding.

Another initiative, titled Restoring Active Memory, or RAM,† launched in 2013, was initially targeted to traumatic brain injuries. For this project, DARPA developed a four-by-four sensor array with ninety-six sensors (not quite as many as Gertrude had) that could home in on the activity of up to three neurons at a time—and amplify it. The amplified signals transmit to a computer that translates them into action. As Sanchez proudly put it, "We can place these [sensors] anywhere in your brain or even in your peripheral nervous system."

When researchers at Wake Forest Baptist Medical Center and the University of Southern California placed the device in volunteers with epilepsy who were having neurosurgery for another reason, they focused on the hippocampus.[5] This target should tell you that their goal was memory improvement, and that's what happened. The volunteers took a working-memory test before and after the placement and showed "up to 37 percent" improvement postimplant. For people with epilepsy, that level of improvement can be therapeutically relevant because seizures are associated with memory loss.

---

* Viewable here: www.youtube.com/watch?v=nvUHDK59Igw.
† A play on "random access memory," the system that stores computer data.

The researchers wanted to pick up signals from a part of the hippocampus, the CA3 region, where neurons fire when new information comes in—that is, when a new memory is being formed. And then they wanted to capture the outgoing signals from another part of the hippocampus, the CA1 region, which delivers up memories. Participants were shown a simple image after instructions to try to remember it, and then asked to pick it out from among a set of images on a screen. While they were encoding a memory of the image and then retrieving the memory to choose the correct image from the set, investigators recorded the activity in CA3 and CA1.

With this information, the scientists developed what they called a MIMO (multi-input/multi-output) model to predict how neuron firing would change from memory encoding (CA3) to memory formation (CA1). They used this information to identify the patterns associated with correct identification and develop "codes" that could be played back. Then came the kind of spooky bit. The study participants completed recall tests while the researchers played back the codes to stimulate their CA1 neurons. It was while under this stimulation from the codes derived from CA3 during memory formation that the volunteers showed the improvement in recall.

That improvement was demonstrated in a short-term memory test, though. To see how the playback stimulus would affect long-term recall, the researchers used a longer delay period (seventy-five minutes, rather than a few) before asking participants to select a previously presented image from among a set of images. Under stimulus to the CA1 with the CA3-derived code, the volunteers showed 35 percent better recall compared with their unstimulated performance. The scientists were quick to emphasize that the code didn't replace the work of the CA1 or implant anything. It just drew the CA1's activity in the right direction for accurate recall.

In other DARPA-funded work at the University of Pennsylvania, investigators used a list of words for a recall test, instead of images. And they recorded from and stimulated the lateral temporal cortex, which underlies word recognition. The participants experienced a 15 percent improvement in memory under stimulation.[6] At his 2017 presentation, Sanchez played a video in which one of the participants has just viewed a list of twelve words and is asked to recall them. Without the stimulation, he's able to remember three of the twelve. The video shows him face-palming in frustration. Next, he's administered the stimulus and recalls all twelve. "I had a good picture in my head," he says. "I could just see." It's like neurofeedback minus the human effort.

Some researchers are working on placing the middleman that delivers the code directly into the brain—implanting it as a microchip that could take information from one area of the cortex, process it, and then relay output to another area. If that happens, we truly will become part robot and can genuinely start referring to our brains as "wired."[7]

In his presentation, Sanchez spun these results into broader applications. Someday, he predicted, we will be able to sit on our couches and participate in a "direct neural interface" with the outside world or with our friends, like Professor X of *X-Men* donning the Cerebro device and communing with the world. Or maybe you're getting older, Sanchez posited, and want to record your memories to revisit them later or share with friends and family. An interface like that could do it, he said.[8]

But what else could be recorded and shared that, perhaps, we don't want shared? Sanchez acknowledged the pitfalls: "We know that these technologies could be used for good or for ill." There are many drawbacks and ethical knots to untie, and DARPA itself has historically created a few ethical tangles.* Toward the end of the chapter, I will discuss some of the ethics that need attention as we face our BCI future. Given the nature of this technology and the currently high barriers to accessing it, the stage is already set for a disconnect between the neuroenhanced haves and have-nots.

DARPA also has funded work on implants in the motor and sensory cortices that can send movement signals to paralyzed limbs. Sanchez showed a video of a research participant who had lost his hand in an electrical accident and then had a target sensor implanted in the nerves of his forearm. When he reached out a virtual hand to touch and open a virtual door, the arm sensor sent a signal to the sensory cortex that the touch was happening. He felt it as real. His reaction? "God, that is so cool."

He's not wrong.

---

* Examples include irradiating servicemen without their knowledge (and obviously without their consent).

This approach, using an individual's brain signatures to produce a "corrective" code for that individual is a step up from brute-force stimulus with brain implants. The latter simply targets the region around the implant with various frequencies, with, as one review author wrote in 2019 (wait for it), "limited success and contradictory results."[9] Those contradictions include disruptions in memory, which is the opposite of the goal.

What remains unclear is if such memory injections could transfer, correcting the memory of one person with the code from another. Researchers have achieved something like this with rats. They take the codes from trained rats that have mastered a task and use them to stimulate untrained rats. The untrained rats, in turn, execute the trained activity right away.[10] In writing about their work, the authors use the phrase "direct donor–recipient manner," which, as Elon Musk might put it, draws us straight into a *Dark Mirror* episode.

DARPA rebooted its RAM program in 2015 as RAM Replay, which is focused on ways to tweak the brain without invading it.[11] After all, the agency's first commitment is to people serving in the US armed forces, which explicitly excludes anyone with epilepsy from enlisting. Scientists typically can't use surgical implants in people who don't have a medical indication for one, so the logical next step was to investigate nonsurgical methods. The same model of picking up brain patterns and replaying them as a stimulus to memory applies, except it's achieved from outside the cranium.

One study relied on memory consolidation during dreaming. Participants navigated a virtual city while awake, listening to sounds researchers played as they did so. The sounds were later replayed (hence RAM Replay) to some participants during their sleep, while a control group didn't receive the replay. When tested later, the participants who experienced sound replay during sleep could make their way through the virtual city about 40 percent faster than those who didn't. DARPA spookily refers to this as "targeted memory reactivation."

DARPA seems to like the term "targeted": they used it again in a 2016 program named Targeted Neuroplasticity Training.* Its aim is to see if

---

* Yes, that abbreviates as TNT.

stimulation of nerves outside the brain, in the peripheral nervous system, can trigger precision neurotransmitter release in the brain and promote new synapse formation—and, presumably, new capacities. This outcome may sound eerie, but it might be better than brain implants. The thing about implants, at least so far, is that they can't stay in for very long, and they carry a lot of risks. The body is not a fan of foreign objects, and the brain is particularly defensive. As a result, these prostheses can trigger re-actions that are potentially quite damaging. That doesn't sound like an acceptable trade-off.

To limit the risks associated with neurosurgical implantation, some groups are working on an injectable stimulator, which hypothetically can *flow* to the target region and then behave as an electrode (aka an "electrically conductive interface") once it settles in. The researchers who developed the flowing electrode have dubbed their creation the Injectrode, which is my favorite neuroscience term of the millennium so far. They've tried it out in rats and, of course, pigs.[12]

In another DARPA-funded study, researchers used a different approach in two patients who had amyotrophic lateral sclerosis and paralysis of their upper limbs. The scientists implanted a stimulator inside a large vein* that drains fluid from the brain and runs alongside the primary motor cortex.[13] In a sense, the procedure is like inserting a stent to prop open a blood vessel, as has been done thousands of times in quick operations for people with heart disease (which is a little less fraught than brain surgery). The device, appropriately called Stentrode, communicates EEG signals wirelessly to a computer. The two participants had to spend time learning how to use eye motion (which sends messages to the motor cortex) to generate EEG signals to move an eye-tracker cursor. Their brief was to learn to move a mouse and click it to zoom or to select options on a screen, all by way of eye motions transmitted through the implant as signals to the computer.

After the training, the two implant recipients were then left "unsuper-vised," meaning left literally to their own devices to achieve some activities of daily living. On their own, they were able to type accurately up to twenty

---

* For the intensely curious and detail-oriented reader, it was the superior sagittal sinus.

characters per minute, and to use the mouse to shop, send text messages, and manage their finances online.

CALLING DEATH INTO QUESTION

The idea that you might be able to record neural signals of your own memories and transmit them to another human being raises the question of raising the dead. Or of immortality. If the products of your brain live on without your body, and if your brain is you, then as long as these signals exist, would you ever truly be gone? It's a modern twist on the age-old belief that if people still have memories of you, you're never truly gone—except in this case, they have your memories too. The memories would exist without your body—but also without your brain.

What about a form of immortality in which your brain keeps on keepin' on, disembodied? *Futurama* and *The Simpsons* have portrayed brain-cyborg mash-ups as characters on their shows—one set in the future (*Futurama*, natch) and the other when it makes forays into its characters' futures (*The Simpsons*). The characters usually are presented as consisting only of heads, as floating in fluid, or as navigating about on a robotic body. Richard Nixon recurs as a character on *Futurama*, his head confined in a Bell jar—which doesn't seem like a desirable real-world outcome in so many ways. In "Future Drama," one of many episodes of *The Simpsons* that peek at the future, Homer eventually just exists as his consciousness on a flash drive that initially is connected to monitors to show his moving image, but that Marge later combines with a robot.

No, we have yet to achieve anything like that with humans. But in 2019, researchers did revivify the brains of some disembodied pigs.[14] Perhaps the most striking aspect of their work (beyond the whole revivification thing) is that the pigs weren't special in any way. They weren't selected for the lab but were reared as meat animals. They weren't killed by careful injections under controlled conditions but instead were slaughtered in a US Department of Agriculture slaughterhouse. And then the brains of three hundred of them were sent to a lab at Yale.

Hours after the animals died, the Yale researchers and their colleagues perfused thirty-two of the brains with a solution they dubbed BrainEx, designed to behave like blood flow to the organ. The liquid contained oxygen and other substances brains need to function, including glucose. The cells of these hours-dead pig brains took up the fuel and began using it, waking up and commencing something like normal function. The cells showed signs of mounting an immune defense, and when researchers took neuron samples, the neurons could fire.* The experiment was conducted at room temperature, or as the researchers put it, under "ex vivo normothermic conditions." Some of the brains stayed in this apparently alive state for thirty-six hours after the perfusion, when the experiments ended, largely for logistical reasons. Perhaps they could have stayed in that partially animated condition for even longer. Brains perfused with a placebo solution or not perfused at all just . . . stayed dead.

The ethical implications of this study are enormous. Some of them relate to using loss of brain function as a marker of death for organ donation, or to the use of human brain tissue in research.[15] If brains that are removed from skulls can be revived after four hours—which the researchers called "an unappreciated capacity" of a large mammalian brain—what are the implications for how we perceive and define medical death? And for what it means to be alive?[16] In an editorial that accompanied the publication of the study, another author group offered what may be the first *Princess Bride* quote ever used in neuroscience literature: "'There's a big difference between mostly dead and all dead. Mostly dead is slightly alive.'"

The editorial authors also suggested that if anyone wants to take the next step and evaluate these Piggenstein-by-perfusion brains for activity, then TMS, rather than EEG, would be the better approach. EEG, they wrote, isn't a reliable indicator of consciousness because it detects signals in people who are unconscious under general anesthesia. But TMS, with its stimulus directed at brain targets and intended to trigger a response, could be one way to see if those apparently alive, perfused brains can respond to a stimulus.

To quote *Wired* journalist Adam Rogers, who covered Elon Musk's Neuralink/Gertrude unveiling, before we can answer these questions, "someone

---

* This happened in spite of the fact that they had deliberately used chemicals in BrainEx to keep neurons from firing, which wanders into the murkiest of ethical territories.

is going to have to know something about how brains work. . . . Neuroscience doesn't even have a consistent theory of consciousness yet."[17]

A LITTLE KNOWLEDGE

Never one to let a little knowledge be a dangerous thing, Musk has pushed forward undaunted, moving on from rats, the occasional sheep, and Gertrude the pig to monkeys. Neuralink Corporation has reported plans to implant devices into a monkey that allow the animal to play "mind pong" with a robot monkey. Musk revealed this work in the most 2021 way possible, giving a private speech on the social app Clubhouse.[18] His project is another version of electrodes on strings, this time a total of about three thousand, keeping tabs on about one thousand neurons in the monkey's brain. Musk reportedly remarked that the animal is "a happy monkey," and that it lives in what he says the USDA inspector called "the nicest monkey facility" they'd reviewed.

It's not necessarily a huge step from monkeys to people in scientific research. In fact, monkeys are often the last preclinical—or prehuman—study animal before researchers advance to people, which is exactly what Elon Musk plans to do next. And through that door, there be dragons.

It's one thing to try to get a meatspace monkey to play ping-pong with a cyborg, but it's another thing entirely to expose people's brains to the risks of being "brainjacked." The reliance on wireless communication widens the vulnerability gap. If that doesn't make you wary, how about the fact that Facebook—*Facebook*—is reportedly working on a BCI that can read brain activity?[19]

Futurists have looked ahead and seen different kinds of implants that might serve different purposes for us, such as a chip for sharing thoughts with other people, or one that turns you into a human abacus.[20] Setting aside the question of why you'd want someone to be able to read your thoughts, even the use of BCIs available today raises some pretty pointed ethical issues. Right now, the idea of using such devices for purposes beyond medical indications looks like a lot of thorns and few roses. Could someone steal our thoughts? Embed their own thoughts in our minds? Control our thoughts?

These used to be questions for science fiction writers to wrestle with, but they are steps away from reality today.

We already face some of these concerns on fronts that are indirectly related to our brains. When we rely on Alexa to run our homes, we're using the environment to offload part of our cognitive burden. This neuroergonomic tactic carries risks, though. We set alerts and appointments on our electronic calendars. We RSVP to an emailed invite requiring us only to click "yes" to automatically make an event show up on synchronized calendars. As we take advantage of these and other memory-saving accommodations, we give something away: our data and our privacy.

Now imagine that these tools become invasive. Monitors are developed that can use EEG to keep track of our cognitive load and, with the detection of certain pattern changes in our brain waves, alert us that we're overdoing it.[21] That seems OK until you consider that an employer might require employees to use this technology to avoid, say, workers' compensation claims related to overlong hours.

A more positive use might involve your car detecting brain waves that suggest inattention, and alerting you. Automotive companies have looked into this possibility. Accident prevention is certainly a desirable outcome, but the trade-off is that someone—or some company, which is a person, according to the US Supreme Court—is collecting your data to use for purposes good or ill. The authors of a 2021 review of the ethics of BCIs wrote, "Consideration of ethical issues will become increasingly important as the technology for artificially injecting various kinds of information into the nervous system improves and enters the mainstream."[22] The idea of injections is scary to a lot of people when they're simply administered in an arm. Injections straight to the brain? Limbic brain says: fear and disgust.

That hasn't stopped quite a few companies from trying to peddle wearable BCIs that they promise will tweak your mood, help you sleep, or just permit you to use EEG on yourself all the time. One such firm, NeuroSky, claims that its EEG sensors will help you "Understand the mind. Unleash its power"—which as you have seen in this book would put you well ahead of the entire field of neuroscience.

These businesses are marketing to other businesses that want to sell their memory-improvement, focus-improvement, sharpness-improvement products to you. And they want to sell you games that you "can play with your mind," which they hope to popularize as "neurogaming." This activity involves a weak form of neurofeedback because it uses brain wave measurements, in a single channel rather than many, and only lets you "move things with your mind" once you emit waves that are associated with "focus and calmness."

So far, the neurogames seem to be a bit of a fizzle, which may be fine because no one's quite sure about the privacy repercussions of serving up your brain wave patterns to total strangers for nonmedical reasons. Other efforts to market straight to consumer have not rocketed off the launchpad either. That's all right for now, because as the authors of a 2020 systematic review noted, "Little research has been done into the intersection of the current state-of-the-art commercial applications of EEG and their ethical concerns."[23] These authors found that in articles discussing ethical issues, the most common themes relate to the balance of safety risks versus benefits, personal agency (such as informed consent and awareness of the risks and benefits), maintaining a personal identity, the effects of enhancement, and data and privacy protections.

To consider one example, the long-term effects of a neurogame that requires the player to wear an EEG-equipped headset are unknown, especially on developing brains. Furthermore, no one knows whether potential side effects would be reversible. If users of brain-training devices (should they prove to work) become reliant on them, what would happen if the devices fail or malfunction?

The authors of the 2020 review concluded, "Most scholars think that neurotechnologies could clearly disrupt the physical and mental integrity of the individual," raising the identity concerns, especially with invasive BCIs. They cited studies showing some unwanted side effects of existing interventions, which show benefits for Parkinson's disease, for example, but also are linked to mania, gambling, hypersexuality, and other behavioral changes. Recipients have reported feeling "like a robot" or "electric doll," or as if they were being handled by "remote control."

More optimistic proponents of such technology think their promise is more enhancing than enharming. One author has even predicted that with the uptake of such developments, humans could advance from the species *Homo sapiens sapiens* (modern humans) to *Homo sapiens technologicus* (cyborg humans),* especially if implants or chips are devised that relay messages from one electrode to another inside our heads.[24] Should that scenario arise, regular humans might coexist with cyborgs, and society would take on some kind of dystopian new world order depending on who had the upper hand . . . or brain.

What if—just stay with me here—one of these interventions with grand claims actually does work someday? You use it, and it changes your brain, so much so that you're no longer even sure you're you. In addition to concerns about having your private thoughts, brain waves, and data about your daily activities captured by unseen entities to use as they wish, there are concerns that such an intrusion could sponge up what makes you you.

These neuroprivacy issues are not trivial,[25] and the concerns are here, now, as more and more wearable devices transmit our health data to invisible entities. No one ever seems ready to address these types of apprehensions before the horse leaves the barn, and brain-excavating tools are no exception. What's different is that some concerned ethicists view brain data as exceptional, more personal and invasive to collect than, say, data on your heart rate. I can't disagree.

BRAIN NET

In a startling study published in 2019, researchers described a brain net that they dubbed, perhaps unimaginatively, BrainNet, a neurogame in which three players send and receive information using only their brains.[26] Two of the players, the Senders, wear EEG caps and have a bird's-eye view of what needs to happen in the game, but they cannot do it themselves. The third player is the Receiver, who lacks the overview but can receive guidance from the Senders. That guidance, which comes in from the EEG caps, triggers a magnetic stimulus in the Receiver's brain.

---

* Of course, unless the cyborg part can trace to genetic variants of some kind, this isn't evolution in any biological sense but rather mere wordplay.

In the game, the action to be taken is a binary: yes or no. To give it the nod, the Senders stare at a part of the computer screen that reads "yes." To decline the action, they stare at the word "no," positioned on the opposite side of the screen. Each of the words flashes a specific number of times, and the EEG registers them differently. The EEG information sent to the Receiver reflects the difference in number of flashes for yes versus no. The Receiver gets the message not by seeing it but by TMS stimulation of the visual cortex in the back of the head, in the occipital lobe. If the signal from the Senders is a "yes," then the Receiver "sees" a bright flash. If it's a "no," the Receiver registers nothing.

The researchers who demonstrated the brain net used a simplified version of a *Tetris*-like game. The yes signal indicated to the Receiver that the block portrayed on the screen should be rotated. A double yes meant that the Senders agreed about what to do. Five different trios successfully managed to play the game, with an average "rotate the block" accuracy of 81.25 percent. The authors also found that Receivers could tell which Sender was more reliable and would follow that player's incoming instruction more readily. The upshot was that the researchers had created a genuine, wired electromagnetic social network of three individual brains that showed typical social behaviors (e.g., trust the reliable one more).*

Miguel Nicolelis, the Musk skeptic who's been deeply immersed in BCI research for years, thinks that no matter how good we get at connecting brains to machines or brains to brains, we'll never achieve a brain completely detached from "substrate": the physical components that cognition affects and is affected by. No Nixon in a Bell jar or Homer downloaded to a robot.

In his 2020 book *The True Creator of Everything*, Nicolelis writes that, sure, we may be able to recapitulate brain activity related to yes/no logic

---

* An eerie (literal) footnote: scientists have also created mice whose neurons can be activated by implanted lights that can be programmed to flash at different frequencies. When the researchers activate lights simultaneously at the same frequency in a pair of mice that are fighting, the animals stop being aggressive and start grooming each other, a cardinal sign of rodent friendship. When two mice experience different frequencies from the implants, they give each other the cold shoulder. The lights act as switches that bypass the natural interactions that produce resonance and brain wave syncing between two individuals and force this syncing artificially (see Yang et al. 2021).

gates, an imitation of the digital machine that everyone thought the brain might be. But he also argues that our brains rely on nondigital processes that cannot be recapitulated. All the information that comes in those bits and bytes alters the physical components of our brains because of how the information acts on them. In response, our neurons fire and move ions around, neurotransmitters ebb and flow, blood swishes past, oxygen comes and goes, and electromagnetic fields are generated.

This is the ecosystem of Planet Brain, not wired and plugged or unplugged, but constantly changing in interaction with special deliveries from within itself and the rest of the body. In turn, the brain alters itself and the rest of the body by what it sends back out. With that response, we interact again with our internal and external environments, generating more information that alters the physical components of our brain. And so it goes on.

We are integrated, body and brain, brain and body, inside and out, from molecules to cells to tissues to organs to the organism to other people. A ripple in one level of the structural hierarchy causes ripples across them all, like the small earthquakes that cascade into a major temblor, or the firings of a single neuron that can lead to an avalanche of signals. As the most consistent elements of our own environments, we each provide one another with information that exerts physical effects in our brains. We are all part of one another's integration.

Nicolelis describes the human brain as an "organic computer." If so, then collectively we are organic brain nets. As illustrated by the results of the Sender-Receiver *Tetris*-like game, we can even exercise social cognition using only messages sent straight from one brain to another. Yet even in that study, the people still had their bodies and used them.

"But the pigs!" you interpolate. Their brains were disembodied, and a perfusion of special brain juice still woke them up! The pigs are intriguing, sure. But as numerous commentaries on that research pointed out, no one demonstrated that the brains were doing the one thing brains are best known for doing: firing neurons in response to inputs or to trigger outputs. They certainly weren't behaving anything like an actual pig, and in the absence of a call-and-response relationship with their environment, they weren't anything like an organic computer. They were tissues reawakened by a carefully concocted mixture of cellular food and drink, but that doesn't mean the BrainEx was an elixir of life.

In the absence of measurable brain waves, no one can say whether these semirevived pig brains, unencumbered by skulls, could have resonated with each other and gotten into synch. But studies of humans point to that possibility, potentially with the behavioral responses our brains generate as mediators. Moving a certain way can serve as a form of synchronicity, as when two people who have never met start spontaneously mimicking each other on the dance floor, or two people raise eyebrows at each other over someone else's remark. This immediate recognition of a resonance based on motor outputs from two brains can be striking, and it is a key part of our social cognition and recognition. How Nixon in a Bell jar could replicate that remains unclear.

Studies relying on every technique you've learned about here—EEG, TMS, MEG, fMRI—demonstrate that people who resonate in a social sense also seem to show synchrony in these measures. Playing music together, kissing, or being on a team can result in a "hyperbrain" network, as the authors of one review called it,[27] or a "brain net." Underlying these synchronous physical acts, which are the output of our brains, is an actual synchrony in the music of our minds.

I noted at the beginning of the book that most brain-improvement peddlers promise you individual improvement. As you can see, the future seems to hold out a similar assurance, with various implants and stimulation devices poised to help you think faster, better, smarter. With one exception— the BrainNet trio playing the *Tetris*-like game—even the hypothetical brain-tailoring tools of the future aren't focused on how tailoring our brains might affect other brains, especially in beneficial ways. And that's too bad.

Consider the meaning of "synchrony." People literally operating on the same wavelengths seem to get into a positive cadence with each other. The brain waves of musicians playing solo will shift into a synchronous rhythm when they start to follow each other in tone and beat. As we have seen, studies show that if people mutually adapt their brain waves to be in sync, then they may also get into sync with each other.

If the DARPA studies are any indication, BCIs are on the horizon that will allow brains to connect virtually, allowing for disembodied social connection among still-embodied brains. What's absent from the discussions is how these connections could be used to promote empathy and perspective-taking—to boost our social cognition—and not just to make us memory champs.

If the future includes tailoring tools to enhance our social cognition, then there may still be hope for us all.

ACKNOWLEDGMENTS

In the spirit of the theme of this book, I could not have written it had it not been for the many, many supportive people in my life. For providing executive function and attentional guidance, I am grateful to my editor at Basic Books, Eric Henney, and to my agent at Janklow & Nesbit, Emma Parry. Thanks to both for being professional, kind, insightful, and patient throughout this process. For providing review and reappraisals, I'd like also to thank Kelley Blewster for her editing skill with the later drafts. On the creativity front, my heartfelt and deepest love and gratitude to W. G. Kunze for providing the art for the first chapters.

While writing this book, I consulted with several experts who were kind enough to provide feedback and endure prolonged interviews. For relieving a cognitive load and making information accessible, I thank Judith Moskowitz, Daniel Simons, Giovanni Sala, Lazar Stankov, Carlene MacMillan, Selena Atasoy, Manoj Doss, Alvaro Pascual-Leone, Lasana Harris, Miguel Nicolelis, Dana Smith, Kevin Bird, Bethany Brookshire, Christie Aschwanden, Jens Foell, and a few unnamed kind souls for their time and brain power. They did their best to keep me from stumbling, and any stumbles that happened anyway are entirely my own doing.

As I approached my book deadline after more than a year at home with my spouse and children—cherished time amid a global tragedy—I needed somewhere to go, during a pandemic, where I could work alone, in the quiet, and truly focus. And that's when Jennifer Myers and her family stepped in and made it possible, gratis, for me to spend two weeks among the pine trees of the Sierra and get the job done. It's one of the most generous acts of friendship I've ever experienced and an exemplar of how social support and good companions can ease burdens. I've expressed my gratitude by way of good beer, but I also want to express it here: Jen, your gracious gift to me is a priceless one, and I will never forget it.

And finally, respect, love, and admiration go to my beloveds, who help ease my cognitive burdens and gladden my heart through their humor, love, sense of community, ease of association, and outdoor companionship. Without them, I'd just be a lonely brain.

NOTES

CHAPTER 1: MEET YOUR BRAIN: THE PLANET

1. Kamp et al. 2012.
2. Gross 1987.
3. Yon et al. 2020.
4. Porter 2015.
5. Assaf et al. 2020.
6. Van den Heuvel et al. 2019.
7. Ardesch et al. 2019.
8. Friederici 2017.
9. Ruminjo and Mekinolov 2008.
10. Restrepo-Martínez et al. 2019.
11. Raichle 2015.
12. Raichle 2015.
13. Alves et al. 2019.
14. Yeshurun et al. 2021.
15. Salehi et al. 2020.
16. Carroll 2020; Raposo Pereira et al. 2019.
17. Herculano-Houzel 2013.
18. "How Big Is a Billion?"
19. Blanding 2017.
20. Jasanoff 2018.

CHAPTER 2: BRAIN TINKERING: TOOLS AND TECHNIQUES

1. Stahl et al. 2018.
2. Kadakia et al. 2021.
3. Biswal et al. 1995.
4. Coenen and Zayachkivska 2013.
5. Coenen and Zayachkivska 2013.

6. İnce et al. 2020.
7. Zeidman et al. 2014.
8. Lisman and Jensen 2013; Horschig et al. 2014.
9. Malik and Amin 2018.
10. Horschig et al. 2014.
11. Malik and Amin 2018.
12. Halgren et al. 2019.
13. Bozinovski and Bozinovska 2019.
14. Brandmeyer and Delorme 2020.
15. Trambaiolli et al. 2021.
16. Jiang et al. 2017.
17. Jwa 2018.
18. Jwa 2018.
19. Jwa 2018.
20. Reinhart et al. 2017.
21. Petrosino et al. 2020; Horschig et al. 2014.
22. Cinel et al. 2019.
23. Neggers et al. 2015.
24. Pugh et al. 2018; Cinel et al. 2019; Wagner et al. 2018; Ramirez-Zamora et al. 2020; Lavazza 2018; Zuk et al. 2018.

CHAPTER 3: GLOBAL COGNITION I: WHY WE'RE DOING IT WRONG

1. Jones et al. 2011.
2. Murphy et al. 2020.
3. Vicentini et al. 2021.
4. Salehi et al. 2020.
5. Bear et al. 2019; Anticevic et al. 2012.

6. DeSerisy et al. 2021.
7. Haier 2014.
8. Ide 1922.
9. Chipman 1929.
10. Tucker 2005.
11. Hambrick et al. 2020.
12. Hambrick et al. 2020.
13. Richardson and Norgate 2014.
14. Duckworth et al. 2011.
15. Cattell 1936.
16. Jaschik 2009.
17. Haier 2014.
18. Risen 2021.
19. Kan et al. 2019.
20. Sripada et al. 2020.
21. De la Fuente et al. 2021.
22. Segal 2012.
23. Deary et al. 2021.
24. Armstrong-Carter et al. 2020.
25. Deary et al. 2021.
26. Young et al. 2018.
27. Richardson and Norgate 2014.
28. Ritchie and Tucker-Drob 2018.
29. Ritchie and Tucker-Drob 2018.
30. Davies et al. 2018.
31. Deary et al. 2021.

CHAPTER 4: GLOBAL COGNITION II: GAME TIME?

1. Roscoe 2021.
2. DeArdo 2020; "17 seasons of ailments," 2017.
3. Hilgard et al. 2019.
4. Sripada et al. 2020; Shen et al. 2018.
5. Stankov and Lee 2020.
6. Mayor et al. 2020.
7. Dresler et al. 2019.
8. Barrett 2020.
9. Deck and Jahedi 2015.
10. Arsalidou et al. 2013.
11. Mayor et al. 2020.
12. Dresler et al. 2019.
13. Brühl et al. 2019.
14. Dresler et al. 2019.
15. Brühl et al. 2019.
16. Cohen et al. 2020.
17. Spinelli et al. 2020.
18. Cinel et al. 2019.
19. Horschig et al. 2014.
20. Dresler et al. 2019.
21. Neggers et al. 2015.
22. Chi and Snyder 2012.
23. Aihara et al. 2017.
24. Salvi et al. 2020.
25. Wertheim et al. 2020.
26. Horne et al. 2021; Lavazza 2018.
27. Luber and Lisanby 2014.
28. Luber and Lisanby 2014.
29. Momi et al. 2020.
30. Luber and Lisanby 2014.
31. Beynel et al. 2020.
32. Richardson and Norgate 2014.
33. Dresler et al. 2019.
34. Oberste et al. 2021.
35. Oberste et al. 2021.
36. Oberste et al. 2021.
37. Liu et al. 2020.
38. Chen et al. 2020.
39. Xiong et al. 2021.
40. Taren et al. 2017.
41. Hilgard et al. 2019.
42. Sala and Gobet 2019.
43. Kühn et al. 2019.
44. Sala and Gobet 2019.
45. Sala and Gobet 2019.
46. Sala and Gobet 2019.
47. Mani et al. 2013.
48. Mani et al. 2013.
49. Selita and Kovas 2019.
50. Perou et al. 2019.
51. Herculano-Houzel 2020.
52. Sala and Gobet 2019.
53. Brühl et al. 2019.
54. Besharov et al. 2011.
55. Conti et al. 2016.
56. Hambrick et al. 2020.
57. Ritchie and Tucker-Drob 2018.
58. Ritchie and Tucker-Drob 2018.
59. Stankov and Lee 2020.
60. Hegelund et al. 2020.

61. Ma 2009.
62. Yeshurun et al. 2021.
63. Stankov and Lee 2020.
64. Stankov 1986.
65. Ybarra et al. 2008.

CHAPTER 5: SOCIAL COGNITION: TAILORING THE BRAINS

1. Taylor and Celiberti 2010.
2. Gibson and Douglas 2018.
3. Esménio et al. 2019.
4. Umbach and Tottenham 2020.
5. Tang et al. 2016.
6. Anderson et al. 2018.
7. Christov-Moore et al. 2020.
8. Peveretou et al. 2020.
9. Christov-Moore et al. 2020.
10. Campos et al. 2019.
11. Leblanc and Ramirez 2020.
12. Brown 2020.
13. Waytz and Gray 2018.
14. Heberlein and Adolphs 2004.
15. Katsumi et al. 2021.
16. Li et al. 2014; Esménio et al. 2019; Che et al. 2014; Christov-Moore et al. 2020; Bilevicius et al. 2018; Fareri et al. 2020; Ho et al. 2021; Katsumi et al. 2021.
17. Yeshurun et al. 2021.
18. Taiwo et al. 2021.
19. Yeshurun et al. 2021.
20. Parkinson et al. 2018.
21. Kaplan et al. 2017.
22. Yeshurun et al. 2021.
23. Fishburn et al. 2018.
24. Kinreich et al. 2017.
25. Eddy 2019.
26. Holt-Lunstead et al. 2010.
27. Leblanc and Ramirez 2020.
28. Vatansever et al. 2017.
29. Jenkins 2019.
30. Spunt et al. 2015.
31. Carroll 2020.
32. Krznaric 2012.
33. Carroll 2020.
34. Buckner et al. 2008.

35. Mar 2018.
36. Dodell-Feder and Tamir 2018.
37. Tamir et al. 2016.
38. Mar 2018.
39. Campos et al. 2019.
40. Ardenghi et al. 2021.
41. Luberto et al. 2018.
42. Trautwein et al. 2020.
43. Arulrajan et al. 2020.
44. Peetz and Grossmann 2020.
45. Katsumi et al. 2021.
46. Yeshurun et al. 2021.
47. Pattee 2020.
48. Peetz and Grossmann 2020.

CHAPTER 6: STRESS AND ANXIETY COGNITION: LIGHTENING THE LOAD

1. Zhang et al. 2017.
2. Moore 2009.
3. Fan et al. 2018.
4. Cui et al. 2011.
5. Associated Press 2009.
6. Wong 2009.
7. Wong 2009.
8. Zhang et al. 2017.
9. Munjuluri et al. 2020; Akiki et al. 2018; Lanius et al. 2020.
10. Nicholson et al. 2020; Joshi et al. 2020.
11. Bauer et al. 2019; Lavretsky and Feldman 2021.
12. Garcia et al. 2020.
13. Qiao et al. 2020.
14. Coan et al. 2017.
15. Epel et al. 2018.
16. Moskowitz et al. 2017.
17. Gandy et al. 2020.
18. Meredith et al. 2020.
19. Garakani et al. 2020.
20. Szigeti et al. 2021.
21. Monson et al. 2020; Wolfson et al. 2020.
22. Mitchell et al. 2021.
23. Wolfson et al. 2020.
24. Preller and Vollenweider 2019.
25. Preller and Vollenweider 2019.
26. Preller and Vollenweider 2019.

27. Young et al. 2019; Hoffman et al. 2019.
28. Sholler et al. 2020.
29. De Brouwer et al. 2019; Wright et al. 2020.
30. Shannon et al. 2019.
31. Bergamaschi et al. 2011; Crippa et al. 2011.
32. Spinella et al. 2021.
33. Hundal et al. 2018.
34. Bonaccorso et al. 2019.
35. Larsen and Shahinas 2020.
36. Epel et al. 2018.
37. Distilled from Hagger et al. 2020.
38. Raymond et al. 2019.
39. Perchtold-Stefan et al. 2020.
40. Włodarczyk et al. 2020.
41. Spurny et al. 2021.
42. Distilled from Włodarczyk et al. 2020.
43. Hartman et al. 2007; Pavón et al. 2020; Brietzke et al. 2018.
44. Rawat et al. 2020.
45. Gzielo et al. 2019.
46. Shadick et al. 2013.
47. Woods et al. 2020.
48. Oró et al. 2021.
49. Gandy et al. 2020.
50. Doering et al. 2013; Perri et al. 2021.
51. Cuijpers, Van Veen, et al. 2020.
52. Cuijpers, Van Veen, et al. 2020.
53. American Psychological Association 2017.
54. Keltner and Bonanno 1997.
55. Munjuluri et al. 2020.
56. Miles et al. 2021.
57. Nelson et al. 2016; Miles et al. 2021.
58. Ungar 2019.

CHAPTER 7: ATTENTION AND MEMORY COGNITION: GAINING FOCUS

1. Vallat et al. 2020.
2. Plailly et al. 2019.
3. Vallat and Ruby 2019.
4. Spanò et al. 2020.
5. Cobb 2020.
6. Leblanc and Ramirez 2020.
7. Mankin and Fried 2020.
8. Ankri et al. 2020.
9. Murphy et al. 2020.
10. Sala and Gobet 2019; Nęcka et al. 2021.
11. Changeux et al. 2021.
12. Johnson et al. 2020.
13. Boran et al. 2019; Wu et al. 2021.
14. Johnson et al. 2020.
15. Boran et al. 2019; Johnson et al. 2020.
16. Changeux et al. 2021.
17. Liu et al. 2019.
18. Schneider et al. 2020.
19. Vatansever et al. 2017.
20. Dubravac and Meier 2020.
21. Dubravac and Meier 2020.
22. Leblanc and Ramirez 2020.
23. Higgins et al. 2021.
24. Higgins et al. 2021.
25. "Eight-time world memory champion Dominic O'Brien: Learn how to learn" 2018; "Mastering your memory Dominic O'Brien" 2019.
26. Dresler et al. 2017.
27. Melby-Lervåg and Hulme 2013; Au et al. 2015.
28. Li et al. 2021; Nęcka et al. 2021.
29. Nęcka et al. 2021.
30. Robillard et al. 2015.
31. Dresler et al. 2017; Wagner et al. 2021.
32. Sala and Gobet 2019.
33. Simons et al. 2016; Sala and Gobet 2019; Li et al. 2021.
34. Bonnechère et al. 2020.
35. Repantis et al. 2010; Schneider et al. 2020; Repantis et al. 2021.
36. Repantis et al. 2021.
37. Schneider et al. 2020; Fond et al. 2015.
38. Schneider et al. 2020.
39. Cameron et al. 2020.
40. Healy 2021.
41. Vann Jones and O'Kelly 2020.
42. Bershad et al. 2019.
43. Family et al. 2020.
44. Hutten et al. 2020.

45. Barrett, Krimmel, et al. 2020.

46. Healy 2021.

47. Yeh et al. 2021.

48. Arvaneh et al. 2019.

49. Yeh et al. 2021; Horschig et al. 2014.

50. Brandmeyer and Delorme 2020.

51. Dubravac and Meier 2020.

52. Ke et al. 2019.

53. Ankri et al. 2020.

54. Wu et al. 2021.

55. Murphy et al. 2020.

56. Gruber and Ranganath 2019.

57. Leblanc and Ramirez 2020.

58. Sims et al. 2011.

59. Ertel et al. 2008; Zhou et al. 2020; Zahodne et al. 2019.

CHAPTER 8: MOOD COGNITION: MANAGING MELANCHOLY

1. Dols 1987.

2. Berryman 2016.

3. Angelino accessed 2021.

4. Jacob et al. 2020.

5. Sparling et al. 2020.

6. Heller and Bagot 2020.

7. Bushman 2019.

8. Huang et al. 2021

9. Ravindran et al. 2016.

10. Kandola et al. 2019.

11. Fan et al. 2020; Davis et al. 2020.

12. Allida et al. 2020.

13. Vollenweider and Preller 2020.

14. Kuypers et al. 2019.

15. Watts et al. 2017.

16. Preller and Vollenweider 2019.

17. Agin-Liebes et al. 2020.

18. Carhart-Harris et al. 2017.

19. Barrett, Doss, et al. 2020.

20. Cormier 2020.

21. Doss et al. 2020.

22. Davis et al. 2020.

23. Carhart-Harris et al. 2021.

24. Abdallah et al. 2021.

25. Kadriu et al. 2021.

26. Krystal et al. 2019.

27. Krystal et al. 2019.

28. De Jesus et al. 2021.

29. McNamara and Almeida 2019.

30. Zhang et al. 2019.

31. Liao et al. 2019.

32. Deane et al. 2021.

33. McPhilemy et al. 2020.

34. Suradom et al. 2020.

35. Thesing, Milaneschi, et al. 2020; Thesing, Lamers, et al. 2020.

36. Carney et al. 2019.

37. Okereke et al. 2020.

38. Selhub 2020.

39. Włodarczyk et al. 2021; Ricci et al. 2020.

40. Włodarczyk et al. 2021.

41. Sani et al. 2018.

42. Trambaiolli et al. 2021.

43. Neggers et al. 2015.

44. Hollon 2020.

45. López-López et al. 2019; Cuijpers, Karyotaki, et al. 2020.

46. Smith et al. 2018.

47. Li et al. 2020.

48. Schuman-Olivier et al. 2020.

49. Schuman-Olivier et al. 2020.

50. Saris et al. 2020.

51. Taylor et al. 2020.

CHAPTER 9: CREATIVE COGNITION: UNLOCKING INNOVATION

1. Orsolini et al. 2018.

2. Williams 2021.

3. Abraham 2018.

4. Abraham 2016.

5. Beaty et al. 2016; Uddin 2021.

6. Carhart-Harris et al. 2016.

7. Pennycook et al. 2020.

8. Carroll 2020.

9. Andrews-Hanna and Grilli 2021.

10. Buckner and DiNicola 2019.

11. Hoel 2020.

12. Beggs and Plenz 2003.

13. Varley et al. 2020.

14. Cocchi et al. 2017.

15. Varley et al. 2020.
16. Atasoy et al. 2016.
17. Neubaeur et al. 2018.
18. Neubauer et al. 2018.
19. Uddin 2021.
20. Xie et al. 2021.
21. Madore et al. 2019.
22. Beaty, Kaufman, et al. 2016.
23. Xie et al. 2021.
24. Beaty et al. 2018.
25. Beaty, Benedek, et al. 2017.
26. Beaty et al. 2018; Kaufman 2018.
27. Bowden and Jung-Beeman 2003.
28. Abraham 2018.
29. Beaty, Benedek, et al. 2017.
30. Abraham 2016.
31. Kaufman 2018.
32. Beaty, Kaufman, et al. 2016.
33. Jacobson 1908.
34. Van Elk et al. 2019.
35. Van Elk et al. 2019.
36. Schmid et al. 2015.
37. Mason et al. 2021.
38. MacLean et al. 2011.
39. Kadriu et al. 2021.
40. Leptourgos et al. 2020.
41. Leptourgos et al. 2020.
42. Carhart-Harris et al. 2016.
43. Nicolelis 2020, p. 76.
44. Kadriu et al. 2021.
45. Mason et al. 2020.
46. Kadriu et al. 2021.
47. Benedek et al. 2017; Benedek and Zöhrer 2020.
48. Hassevoort et al. 2020.
49. Conner et al. 2015.
50. Colzato et al. 2015.
51. Colzato et al. 2013.
52. Frith et al. 2019.
53. E.g., Agnoli et al. 2018.
54. Rosen et al. 2016.
55. Henriksen et al. 2020.
56. Lebuda et al. 2016.
57. Agnoli et al. 2018.
58. Preiss and Cosmelli 2017.
59. Laland 2017.

CHAPTER 10: A FREAKY FUTURE: FAR OUT BUT NOT FAR AWAY

1. Tobin et al. 2019.
2. Andrews-Hanna and Grilli 2021.
3. Musk and Neuralink 2019.
4. Brodwin and Robbins 2020.
5. Hampson et al. 2018.
6. Ezzyat et al. 2018.
7. Mazurek and Schieber 2021.
8. Golembiewski 2020.
9. Cutsuridis 2019.
10. Deadwyler et al. 2013.
11. Sanchez and Miranda 2019.
12. Trevathan et al. 2019.
13. Oxley et al. 2021.
14. Vrselja et al. 2019.
15. Youngner and Hyun 2019.
16. Farahany et al. 2019.
17. Rogers 2020.
18. Purslow 2021.
19. Golembiewski 2020.
20. Golembiewski 2020.
21. Lopez et al. 2020.
22. Mazurek and Schieber 2021.
23. Lopez et al. 2020.
24. Zehr et al. 2015.
25. Minielly et al. 2020.
26. Jiang et al. 2019.
27. Mende and Schmidt 2021.

SELECTED BIBLIOGRAPHY

"17 seasons of ailments: Tom Brady's entire reported injury history." ESPN, May 17, 2017. www.espn.com/nfl/story/_/id/19402769/tom-brady-entire-reported-injury-history-nfl.

Abdallah, Chadi G., et al. "A robust and reproducible connectome fingerprint of ketamine is highly associated with the connectomic signature of antidepressants." *Neuropsychopharmacology: Official Publication of the American College of Neuropsychopharmacology* vol. 46, 2 (2021): 478–485. DOI: 10.1038/s41386-020-00864-9.

Abraham, Anna. *The Neuroscience of Creativity*. Cambridge Fundamentals of Neuroscience in Psychology. Cambridge, UK: Cambridge University Press, 2018.

Abraham, Anna. "Gender and creativity: An overview of psychological and neuroscientific literature." *Brain Imaging and Behavior* vol. 10, 2 (2016): 609–618. DOI:10.1007/s11682-015-9410-8.

Agin-Liebes, Gabrielle I., et al. "Long-term follow-up of psilocybin-assisted psychotherapy for psychiatric and existential distress in patients with life-threatening cancer." *Journal of Psychopharmacology* vol. 34, 2 (2020): 155–166. DOI: 10.1177/0269881119897615.

Agnoli, Sergio, et al. "Enhancing creative cognition with a rapid right-parietal neurofeedback procedure." *Neuropsychologia* vol. 118, Pt A (2018): 99–106. DOI: 10.1016/j.neuropsychologia.2018.02.015

Aihara, Takatsugu, et al. "Anodal transcranial direct current stimulation of the right anterior temporal lobe did not significantly affect verbal insight." *PloS One* vol. 12, 9 (2017). DOI: 10.1371/journal.pone.0184749.

Akiki, Teddy J., et al. "Default mode network abnormalities in posttraumatic stress disorder: A novel network-restricted topology approach." *NeuroImage* vol. 176 (2018): 489–498. DOI: 10.1016/j.neuroimage.2018.05.005.

Allida, S., et al. "Pharmacological, psychological, and non-invasive brain stimulation interventions for treating depression after stroke." Cochrane Database of Systematic Reviews, no. 1 (2020). DOI: 10.1002/14651858.CD003437.pub4.

Alves, Pedro Nascimento, et al. "An improved neuroanatomical model of the default-mode network reconciles previous neuroimaging and neuropathological findings." *Communications Biology* vol. 2, 370 (2019). DOI: 10.1038/s42003-019-0611-3.

American Psychological Association. *Clinical Practice Guideline for the Treatment of PTSD*. 2017. www.apa.org/ptsd-guideline/ptsd.pdf.

Anderson, Nathaniel E., et al. "Psychopathic traits associated with abnormal hemodynamic activity in salience and default mode networks during auditory oddball task."

Cognitive, Affective and Behavioral Neuroscience vol. 18, 3 (2018): 564–580. DOI: 10.3758 /s13415-018-0588-2.

Andrews-Hanna, Jessica R., and Matthew D. Grilli. "Mapping the imaginative mind: Charting new paths forward." *Current Directions in Psychological Science* vol. 30, 1 (2021): 82–89. DOI:10.1177/0963721420980753.

Andrews-Hanna, Jessica R., et al. "The default network and self-generated thought: Component processes, dynamic control, and clinical relevance." *Annals of the New York Academy of Sciences* vol. 1316, 1 (2014): 29–52. DOI: 10.1111/nyas.12360.

Angelino, Andrew F. "Depression: What you need to know as you age." Johns Hopkins Medicine. Accessed May 5, 2021. www.hopkinsmedicine.org/health/conditions-and-diseases /depression-what-you-need-to-know-as-you-age.

Ankri, Yael L. E., et al. "The effects of stress and transcranial direct current stimulation (tDCS) on working memory: A randomized controlled trial." *Cognitive, Affective and Behavioral Neuroscience* vol. 20, 1 (2020): 103–114. DOI: 10.3758/s13415-019-00755-7.

Anticevic, Alan, et al. "The role of default network deactivation in cognition and disease." *Trends in Cognitive Sciences* vol. 16, 12 (2012): 584–592. DOI: 10.1016/j.tics.2012.10.008.

Ardenghi, Stefano, et al. "An exploratory cross-sectional study on the relationship between dispositional mindfulness and empathy in undergraduate medical students." *Teaching and Learning in Medicine* vol. 33, 2 (2021): 154–163. DOI: 10.1080/10401334.2020 .1813582.

Ardesch, Dirk Jan, et al. "Evolutionary expansion of connectivity between multimodal association areas in the human brain compared with chimpanzees." *Proceedings of the National Academy of Sciences of the United States of America* vol. 116, 14 (2019): 7101–7106. DOI: 10.1073/pnas.1818512116.

Armstrong-Carter, Emma, et al. "The earliest origins of genetic nurture: The prenatal environment mediates the association between maternal genetics and child development." *Psychological Science* vol. 31, 7 (2020): 781–791. DOI:10.1177/0956797620917209.

Arsalidou, Marie, et al. "A balancing act of the brain: Activations and deactivations driven by cognitive load." *Brain and Behavior* vol. 3, 3 (2013): 273–285. DOI: 10.1002/brb3.128.

Arulrajan, Sithhipratha, et al. "Response to: An exploratory cross-sectional study on the relationship between dispositional mindfulness and empathy in undergraduate medical students." *Medical Education Online* vol. 25, 1 (2020). DOI: 10.1080/10872981.2020.1826112.

Arvaneh, Mahnaz, et al. "A P300-based brain-computer interface for improving attention." *Frontiers in Human Neuroscience* vol. 12, 524 (2019). DOI: 10.3389/fnhum.2018.00524.

Assaf, Yaniv, et al. "Conservation of brain connectivity and wiring across the mammalian class." *Nature Neuroscience* vol. 23, 7 (2020): 805–808. DOI: 10.1038/s41593 -020-0641-7.

Associated Press. "Sichuan earthquake killed more than 5,000 pupils, says China." *The Guardian*, May 7, 2009. www.theguardian.com/world/2009/may/07/china-quake -pupils-death-toll.

Atasoy, Selen, et al. "Human brain networks function in connectome-specific harmonic waves." *Nature Communications* vol. 7 (2016). DOI: 10.1038/ncomms10340.

Au, Jacky, et al. "Improving fluid intelligence with training on working memory: A meta-analysis." *Psychonomic Bulletin and Review* vol. 22, 2 (2015): 366–377. DOI: 10.3758 /s13423-014-0699-x.

Barrett, Frederick S., M. A. Doss, et al. "Emotions and brain function are altered up to one month after a single high dose of psilocybin." *Scientific Reports* vol. 10 (2020). DOI: 10.1038/s41598-020-59282-y.

Barrett, Frederick S., S. A. Krimmel, et al. "Psilocybin acutely alters the functional connectivity of the claustrum with brain networks that support perception, memory, and attention." *NeuroImage* vol. 218 (2020). DOI: 10.1016/j.neuroimage.2020.116980.

Barrett, Lisa F. *Seven and a Half Lessons About the Brain.* Boston: Houghton Mifflin Harcourt, 2020.

Bauer, C. C. C., et al. "From state-to-trait meditation: Reconfiguration of central executive and default mode networks." *eNeuro* vol. 6, 6 (2019). DOI: 10.1523/ENEURO.0335-18.2019.

Bear, Joshua J., et al. "The epileptic network and cognition: What functional connectivity is teaching us about the childhood epilepsies." *Epilepsia* vol. 60, 8 (2019): 1491–1507. DOI: 10.1111/epi.16098.

Beaty, Roger E., M. Benedick, et al. "Creative cognition and brain network dynamics." *Trends in Cognitive Sciences* vol. 20, 2 (2016): 87–95. DOI: 10.1016/j.tics.2015.10.004.

Beaty, Roger E., S. B. Kaufman, et al. "Personality and complex brain networks: The role of openness to experience in default network efficiency." *Human Brain Mapping* vol. 37, 2 (2016): 773–779. DOI: 10.1002/hbm.23065.

Beaty, Roger E., et al. "Robust prediction of individual creative ability from brain functional connectivity." *Proceedings of the National Academy of Sciences of the United States of America* vol. 115, 5 (2018): 1087–1092. DOI: 10.1073/pnas.1713532115.

Beggs, John M., and Dietmar Plenz. "Neuronal avalanches in neocortical circuits." *Journal of Neuroscience: The Official Journal of the Society for Neuroscience* vol. 23, 35 (2003): 11167–11177. DOI: 10.1523/JNEUROSCI.23-35-11167.2003.

Benedek, Mathias, et al. "Creativity on tap? Effects of alcohol intoxication on creative cognition." *Consciousness and Cognition* vol. 56 (2017): 128–134. DOI: 10.1016/j.concog.2017.06.020.

Benedek, Mathias, and Lena Zöhrer. "Creativity on tap 2: Investigating dose effects of alcohol on cognitive control and creative cognition." *Consciousness and Cognition* vol. 83 (2020). DOI: 10.1016/j.concog.2020.102972.

Bergamaschi, Mateus M., et al. "Cannabidiol reduces the anxiety induced by simulated public speaking in treatment-naïve social phobia patients." *Neuropsychopharmacology: Official Publication of the American College of Neuropsychopharmacology* vol. 36, 6 (2011): 1219–1226. DOI: 10.1038/npp.2011.6.

Berryman, Sylvia. "Democritus." Stanford Encyclopedia of Philosophy (Winter 2016 Edition), Edward N. Zalta (ed.). December 2, 2016. https://plato.stanford.edu/archives/win2016/entries/democritus/.

Bershad, Anya K., et al. "Acute subjective and behavioral effects of microdoses of lysergic acid diethylamide in healthy human volunteers." *Biological Psychiatry* vol. 86, 10 (2019): 792–800. DOI: 10.1016/j.biopsych.2019.05.019.

Besharov, Douglas J., et al. *The High/Scope Perry Preschool Project.* University of Maryland School of Public Policy, Welfare Reform Academy. September 2011. www.welfareacademy.org/pubs/early_education/pdfs/Besharov_ECE%20assessments_Perry_Preschool.pdf.

Beynel, Lysianne, et al. "Structural controllability predicts functional patterns and brain stimulation benefits associated with working memory." *Journal of Neuroscience: The*

Official Journal of the Society for Neuroscience vol. 40, 35 (2020): 6770–6778. DOI: 10.1523/JNEUROSCI.0531-20.2020.

Bilevicius, Elena, et al. "Trait emotional empathy and resting state functional connectivity in default mode, salience, and central executive networks." *Brain Sciences* vol. 8, 7 (2018). DOI: 10.3390/brainsci8070128.

Biswal, B., et al. "Functional connectivity in the motor cortex of resting human brain using echo-planar MRI." *Magnetic Resonance in Medicine* vol. 34, 4 (1995): 537–541. DOI: 10.1002/mrm.1910340409.

Blanding, Michael. "Brainiac: With her innovative 'brain soup,' Suzana Herculano-Houzel is changing neuroscience one species at a time." *Vanderbilt News*, September 2017. https://news.vanderbilt.edu/2017/09/07/brainiac-with-her-innovative-brain-soup-suzana-herculano-houzel-is-changing-neuroscience-one-species-at-a-time/.

Bonaccorso, Stefania, et al. "Cannabidiol (CBD) use in psychiatric disorders: A systematic review." *Neurotoxicology* vol. 74 (2019): 282–298. DOI: 10.1016/j.neuro.2019.08.002.

Bonnechère, Bruno, et al. "The use of commercial computerised cognitive games in older adults: A meta-analysis." *Scientific Reports* vol. 10, 1 (2020). DOI: 10.1038/s41598-020-72281-3.

Boran, Ece, et al. "Persistent hippocampal neural firing and hippocampal-cortical coupling predict verbal working memory load." *Science Advances* vol. 5, 3 (2019). DOI: 10.1126/sciadv.aav3687.

Bowden, Edward M., and Mark Jung-Beeman. "Normative data for 144 compound remote associate problems." *Behavior Research Methods, Instruments, and Computers: A Journal of the Psychonomic Society* vol. 35, 4 (2003): 634–639. DOI: 10.3758/bf03195543.

Bozinovski, Stevo, and Liljana Bozinovska. "Brain-computer interface in Europe: The thirtieth anniversary." *Automatika* vol. 60, 1 (2019): 36–47. DOI: 10.1080/00051144.2019.1570644.

Brandmeyer, Tracy, and Arnaud Delorme. "Closed-loop frontal midlineθ neurofeedback: A novel approach for training focused-attention meditation." *Frontiers in Human Neuroscience* vol. 14 (2020). DOI: 10.3389/fnhum.2020.00246.

Brietzke, Elisa, et al. "Ketogenic diet as a metabolic therapy for mood disorders: Evidence and developments." *Neuroscience and Biobehavioral Reviews* vol. 94 (2018): 11–16. DOI:10.1016/j.neubiorev.2018.07.020.

Brodwin, Erin, and Rebecca Robbins. "As Elon Musk's Neuralink prepares to draw back the curtain, ex-employees describe rushed timelines clashing with science's slow pace." *STAT*, August 25, 2020. www.statnews.com/2020/08/25/elon-musk-neuralink-update-brain-machine-implants/.

Brown, Steven. "The 'who' system of the human brain: A system for social cognition about the self and others." *Frontiers in Human Neuroscience* vol. 14 (2020). DOI: 10.3389/fnhum.2020.00224.

Brühl, Annette B., et al. "Neuroethical issues in cognitive enhancement: Modafinil as the example of a workplace drug?" *Brain and Neuroscience Advances* vol. 3 (2019). DOI: 10.1177/2398212818816018.

Buckner, Randy L., et al. "The brain's default network: Anatomy, function, and relevance to disease." *Annals of the New York Academy of Sciences* vol. 1124 (2008): 1–38. DOI: 10.1196/annals.1440.011.

Buckner, Randy L., and Lauren M. DiNicola. "The brain's default network: Updated anatomy, physiology and evolving insights." *Nature Reviews: Neuroscience* vol. 20, 10 (2019): 593–608. DOI: 10.1038/s41583-019-0212-7.

Buckner, Randy L., and Fenna M. Krienen. "The evolution of distributed association networks in the human brain." *Trends in Cognitive Sciences* vol. 17, 12 (2013): 648–665. DOI: 10.1016/j.tics.2013.09.017.

Bushman, Barbara A. "Physical activity and depression." *ACSM's Health and Fitness Journal* vol. 23, 5 (2019): 9–14.

Cameron, Lindsay P., et al. "Psychedelic microdosing: Prevalence and subjective effects." *Journal of Psychoactive Drugs* vol. 52, 2 (2020): 113–122. DOI: 10.1080/02791072.2020.1718250.

Campos, Daniel, et al. "Exploring the role of meditation and dispositional mindfulness on social cognition domains: A controlled study." *Frontiers in Psychology* vol. 10 (2019). DOI: 10.3389/fpsyg.2019.00809.

Carhart-Harris, R. L., et al. "Psilocybin with psychological support for treatment-resistant depression: Six-month follow-up." *Psychopharmacology* vol. 235, 2 (2018): 399–408. DOI: 10.1007/s00213-017-4771-x.

Carhart-Harris, Robin L., et al. "Neural correlates of the LSD experience revealed by multimodal neuroimaging." *Proceedings of the National Academy of Sciences of the United States of America* vol. 113, 17 (2016): 4853–4658. DOI: 10.1073/pnas.1518377113.

Carhart-Harris, Robin L., et al. "Psilocybin for treatment-resistant depression: fMRI-measured brain mechanisms." *Scientific Reports* vol. 7 (2017). DOI: 10.1038/s41598-017-13282-7.

Carhart-Harris, Robin, et al. "Trial of psilocybin versus escitalopram for depression." *New England Journal of Medicine* vol. 384, 15 (2021): 1402–1411. DOI:10.1056/NEJMoa2032994.

Carney, Robert M., et al. "A randomized placebo-controlled trial of omega-3 and sertraline in depressed patients with or at risk for coronary heart disease." *Journal of Clinical Psychiatry* vol. 80, 4 (2019). DOI: 10.4088/JCP.19m12742.

Carroll, J. "Imagination, the brain's default mode network, and imaginative verbal artifacts." In *Evolutionary Perspectives on Imaginative Culture*, edited by J. Carroll et al. Cham, Switz.: Springer, 2020. DOI: 10.1007/978-3-030-46190-4_2.

Cattell, R. B. "Is national intelligence declining?" *Eugenics Review* vol. 28, 3 (1936): 181–203.

Changeux, Jean-Pierre, et al. "A connectomic hypothesis for the hominization of the brain." *Cerebral Cortex* vol. 31, 5 (2021): 2425–2449. DOI: 10.1093/cercor/bhaa365.

Che, Xianwei, et al. "Synchronous activation within the default mode network correlates with perceived social support." *Neuropsychologia* vol. 63 (2014): 26–33. DOI: 10.1016/j.neuropsychologia.2014.07.035.

Chen, Feng-Tzu, et al. "Effects of exercise training interventions on executive function in older adults: A systematic review and meta-analysis." *Sports Medicine* vol. 50, 8 (2020): 1451–1467. DOI: 10.1007/s40279-020-01292-x.

Chi, Richard P., and Allan W. Snyder. "Brain stimulation enables the solution of an inherently difficult problem." *Neuroscience Letters* vol. 515, 2 (2012): 121–124. DOI: 10.1016/j.neulet.2012.03.012.

Chipman, Catherine E. "The constancy of the intelligence quotient of mental defectives." *Psychological Clinic* vol. 18, 3–4 (1929): 103–111.

Christov-Moore, Leonardo, et al. "Predicting empathy from resting state brain connectivity: A multivariate approach." *Frontiers in Integrative Neuroscience* vol. 14 (2020). DOI: 10.3389/fnint.2020.00003.

Cinel, Caterina, et al. "Neurotechnologies for human cognitive augmentation: Current state of the art and future prospects." *Frontiers in Human Neuroscience* vol. 13 (2019). DOI: 10.3389/fnhum.2019.00013.

Coan, James A., et al. "Relationship status and perceived support in the social regulation of neural responses to threat." *Social Cognitive and Affective Neuroscience* vol. 12, 10 (2017): 1574–1583. DOI: 10.1093/scan/nsx091.

Cobb, Matthew. *The Idea of the Brain: The Past and Future of Neuroscience.* New York: Basic Books, 2020.

Cocchi, Luca, et al. "Criticality in the brain: A synthesis of neurobiology, models and cognition." *Progress in Neurobiology* vol. 158 (2017): 132–152. DOI: 10.1016/j.pneurobio.2017.07.002.

Coenen, Anton, and Oksana Zayachkivska. "Adolf Beck: A pioneer in electroencephalography in between Richard Caton and Hans Berger." *Advances in Cognitive Psychology* vol. 9, 4 (2013): 216–221. DOI: 10.2478/v10053-008-0148-3.

Cohen, Pieter, et al. "Five unapproved drugs found in cognitive enhancement supplements." *Neurology Clinical Practice* (2020). DOI: 10.1212/CPJ.0000000000000960.

Colzato, Lorenza S., et al. "Food for creativity: Tyrosine promotes deep thinking." *Psychological Research* vol. 79, 5 (2015): 709–714. DOI: 10.1007/s00426-014-0610-4.

Colzato, Lorenza, et al. "The impact of physical exercise on convergent and divergent thinking." *Frontiers in Human Neuroscience* vol. 7, 824 (2013). DOI: 10.3389/fnhum.2013.00824.

Conner, Tamlin S., et al. "On carrots and curiosity: Eating fruit and vegetables is associated with greater flourishing in daily life." *British Journal of Health Psychology* vol. 20, 2 (2015): 413–427. DOI: 10.1111/bjhp.12113.

Conti, Gabriella, et al. "The effects of two influential early childhood interventions on health and healthy behaviour." *Economic Journal* vol. 126, 596 (2016): F28–F65. DOI: 10.1111/ecoj.12420.

Cormier, Zoe. "Psilocybin treatment for mental health gets legal framework." *Scientific American.* December 1, 2020. www.scientificamerican.com/article/psilocybin-treatment-for-mental-health-gets-legal-framework/.

Crippa, José Alexandre S., et al. "Neural basis of anxiolytic effects of cannabidiol (CBD) in generalized social anxiety disorder: A preliminary report." *Journal of Psychopharmacology* vol. 25, 1 (2011): 121–130. DOI: 10.1177/0269881110379283.

Cui, P., et al. "The Wenchuan earthquake (May 12, 2008), Sichuan Province, China, and resulting geohazards." *Natural Hazards* 56 (2011): 19–36. DOI: 10.1007/s11069-009-9392-1.

Cuijpers, Pim, E. Karyotaki, et al. "The effects of fifteen evidence-supported therapies for adult depression: A meta-analytic review." *Psychotherapy Research: Journal of the Society for Psychotherapy Research* vol. 30, 3 (2020): 279–293. DOI: 10.1080/10503307.2019.1649732.

Cuijpers, Pim, S. C. Van Veen, et al. "Eye movement desensitization and reprocessing for mental health problems: A systematic review and meta-analysis." *Cognitive Behaviour Therapy* vol. 49, 3 (2020): 165–180. DOI: 10.1080/16506073.2019.170 3801.

Cutsuridis, Vassilis. "Memory prosthesis: Is it time for a deep neuromimetic computing approach?" *Frontiers in Neuroscience* vol. 13 (2019). DOI: 10.3389/fnins.2019.00667.

Davies, Neil M., et al. "The causal effects of education on health outcomes in the UK Biobank." *Nature Human Behaviour* vol. 2, 2 (2018): 117–125. DOI: 10.1038 /s41562-017-0279-y.

Davis, Alan K., et al. "Effects of psilocybin-assisted therapy on major depressive disorder: A randomized clinical trial." *JAMA Psychiatry*, e203285 (2020). DOI: 10.1001 /jamapsychiatry.2020.3285.

De Brouwer, Geoffrey, et al. "A critical inquiry into marble-burying as a preclinical screening paradigm of relevance for anxiety and obsessive-compulsive disorder: Mapping the way forward." *Cognitive, Affective and Behavioral Neuroscience* vol. 19, 1 (2019): 1–39. DOI: 10.3758/s13415-018-00653-4.

De Jesus, O., et al. "Neuromodulation surgery for psychiatric disorders." *StatPearls*. Updated February 7, 2021. www.ncbi.nlm.nih.gov/books/NBK482366/.

De la Fuente, Javier, et al. "A general dimension of genetic sharing across diverse cognitive traits inferred from molecular data." *Nature Human Behaviour* vol. 5, 1 (2021): 49–58. DOI: 10.1038/s41562-020-00936-2.

Deadwyler, Sam A., et al. "Donor/recipient enhancement of memory in rat hippocampus." *Frontiers in Systems Neuroscience* vol. 7 (2013). DOI: 10.3389/fnsys.2013.00120.

Deane, Katherine H. O., et al. "Omega-3 and polyunsaturated fat for prevention of depression and anxiety symptoms: Systematic review and meta-analysis of randomised trials." *British Journal of Psychiatry: The Journal of Mental Science* vol. 218, 3 (2021): 135–142. DOI: 10.1192/bjp.2019.234.

DeArdo, Bryan. "Tom Brady, Ben Roethlisberger vying for NFL's unwanted QB record during 2020 season." CBS Sports. November 11, 2020. www.cbssports.com/nfl /news/tom-brady-ben-roethlisberger-vying-for-nfls-unwanted-qb-record-during-2020 -season/.

Deary, Ian J., et al. "Genetic variation, brain, and intelligence differences." *Molecular Psychiatry* (2021). DOI: 10.1038/s41380-021-01027-y.

Deck, Cary, and Salar Jahedi. "The effect of cognitive load on economic decision making: A survey and new experiments." *European Economic Review* vol. 78 (2015): 97–119.

DeSerisy, Mariah, et al. "Frontoparietal and default mode network connectivity varies with age and intelligence." *Developmental Cognitive Neuroscience* vol. 48 (2021): 100928. DOI: 10.1016/j.dcn.2021.100928

Dieter, Schmidt. "Letter to the editor re: Zeidman LA, Stone J, Kondziella D. 'New revelations about Hans Berger, father of the electroencephalogram (EEG), and his ties to the Third Reich.'" *Journal of Child Neurology* vol. 32, 7 (2017): 680–681. DOI: 10.1177/0883073817696813.

Dodell-Feder, David, and Diana I. Tamir. "Fiction reading has a small positive impact on social cognition: A meta-analysis." *Journal of Experimental Psychology* vol. 147, 11 (2018): 1713–1727. DOI:10.1037/xge0000395.

Dodell-Feder, David, et al. "Social impairment in schizophrenia: New approaches for treating a persistent problem." *Current Opinion in Psychiatry* vol. 28, 3 (2016): 236–242. DOI: 10.1097/YCO.0000000000000154.

Doering, Stephan, et al. "Efficacy of a trauma-focused treatment approach for dental phobia: A randomized clinical trial." *European Journal of Oral Sciences* vol. 121, 6 (2013): 584–593. DOI:10.1111/eos.12090.

Dols, M. W. "Insanity and its treatment in Islamic society." *Medical History* vol. 31, 1 (1987): 1–14. DOI: 10.1017/s0025727300046287.

Doss, Manoj K., et al. "The acute effects of the atypical dissociative hallucinogen salvinorin A on functional connectivity in the human brain." *Scientific Reports* vol. 10 (2020): DOI: 10.1038/s41598-020-73216-8.

Dresler, Martin, et al. "Hacking the brain: Dimensions of cognitive enhancement." *ACS Chemical Neuroscience* vol. 10, 3 (2019): 1137–1148. DOI: 10.1021/acschemneuro.8b00571.

Dresler, Martin, et al. "Mnemonic training reshapes brain networks to support superior memory." *Neuron* vol. 93, 5 (2017): 1227–1235. DOI: 10.1016/j.neuron.2017.02.003.

Dubravac, Mirela, and Beat Meier. "Stimulating the parietal cortex by transcranial direct current stimulation (tDCS): No effects on attention and memory." *AIMS Neuroscience* vol. 8, 1 (2020): 33–46. DOI: 10.3934/Neuroscience.2021002.

Duckworth, Angela Lee, et al. "Role of test motivation in intelligence testing." *Proceedings of the National Academy of Sciences of the United States of America* vol. 108, 19 (2011): 7716–7720. DOI: 10.1073/pnas.1018601108.

Eddy, Clare M. "What do you have in mind? Measures to assess mental state reasoning in neuropsychiatric populations." *Frontiers in Psychiatry* vol. 10 (2019). DOI: 10.3389/fpsyt.2019.00425.

"Eight-time world memory champion Dominic O'Brien: Learn how to learn." YouTube video. June 14, 2018. www.youtube.com/watch?v=ACw5YVgg4lc.

Epel, Elissa S., et al. "More than a feeling: A unified view of stress measurement for population science." *Frontiers in Neuroendocrinology* vol. 49 (2018): 146–169. DOI: 10.1016/j.yfrne.2018.03.001.

Ertel, Karen A., et al. "Effects of social integration on preserving memory function in a nationally representative US elderly population." *American Journal of Public Health* vol. 98, 7 (2008): 1215–1220. DOI: 10.2105/AJPH.2007.113654.

Esménio, Sofia, et al. "Using resting-state DMN effective connectivity to characterize the neurofunctional architecture of empathy." *Scientific Reports* vol. 9, 1 (2019). DOI: 10.1038/s41598-019-38801-6.

Ezzyat, Youssef, et al. "Closed-loop stimulation of temporal cortex rescues functional networks and improves memory." *Nature Communications* vol. 9, 1 (2018). DOI: 10.1038/s41467-017-02753-0.

Family, Neiloufar, et al. "Safety, tolerability, pharmacokinetics, and pharmacodynamics of low dose lysergic acid diethylamide (LSD) in healthy older volunteers." *Psychopharmacology* vol. 237, 3 (2020): 841–853. DOI: 10.1007/s00213-019-05417-7.

Fan, Siyan, et al. "Pretreatment brain connectome fingerprint predicts treatment response in major depressive disorder." *Chronic Stress* vol. 4 (2020). DOI: 10.1177/2470547020984726.

Fan, Xuanmei, et al. "What we have learned from the 2008 Wenchuan earthquake and its aftermath: A decade of research and challenges." *Engineering Geology* vol. 241 (2018): 25–32.

Farahany, Nita A., et al. "Part-revived pig brains raise slew of ethical quandaries." *Nature* vol. 568, 7752 (2019): 299–302. DOI: 10.1038/d41586-019-01168-9.

Fareri, Dominic S., et al. "The influence of relationship closeness on default-mode network connectivity during social interactions." *Social Cognitive and Affective Neuroscience* vol. 15, 3 (2020): 261–271. DOI: 10.1093/scan/nsaa031.

Fishburn, Frank A., et al. "Putting our heads together: Interpersonal neural synchronization as a biological mechanism for shared intentionality." *Social Cognitive and Affective Neuroscience* vol. 13, 8 (2018): 841–849. DOI:10.1093/scan/nsy060.

Fond, Guillaume, et al. "Innovative mechanisms of action for pharmaceutical cognitive enhancement: A systematic review." *Psychiatry Research* vol. 229, 1–2 (2015): 12–20. DOI: 10.1016/j.psychres.2015.07.006.

Friederici, Angela D. "Evolution of the neural language network." *Psychonomic Bulletin and Review* vol. 24, 1 (2017): 41–47. DOI: 10.3758/s13423-016-1090-x.

Frith, Emily, et al. "Systematic review of the proposed associations between physical exercise and creative thinking." *Europe's Journal of Psychology* vol. 15, 4 (2019): 858–877. DOI:10.5964/ejop.v15i4.1773.

Gandy, Sam, et al. "The potential synergistic effects between psychedelic administration and nature contact for the improvement of mental health." *Health Psychology Open* vol. 7, 2 (2020). DOI: 10.1177/2055102920978123.

Garakani, Amir, et al. "Pharmacotherapy of anxiety disorders: Current and emerging treatment options." *Frontiers in Psychiatry* vol. 11 (2020). DOI:10.3389/fpsyt.2020.595584.

Garcia, Sarah, et al. "tDCS as a treatment for anxiety and related cognitive deficits." *International Journal of Psychophysiology: Official Journal of the International Organization of Psychophysiology* vol. 158 (2020): 172–177. DOI: 10.1016/j.ijpsycho.2020.10.006.

Gibson, Margaret, and Patty Douglas. "Disturbing behaviours: Ole Ivar Lovaas and the queer history of autism science." *Catalyst: Feminism, Theory, Technoscience* vol. 4, 2 (2018). DOI: 10.28968/cftt.v4i2.29579.

Golembiewski, Lauren. "Are you ready for tech that connects to your brain?" *Harvard Business Review*, September 28, 2020. https://hbr.org/2020/09/are-you-ready-for-tech-that-connects-to-your-brain.

Gross, Charles G. "Neuroscience, Early history of." In *Encyclopedia of Neuroscience*, edited by G. Adelman, 843–847. Basel, Switz.: Birkhäuser, 1987.

Gruber, Matthias J., and Charan Ranganath. "How curiosity enhances hippocampus-dependent memory: The prediction, appraisal, curiosity, and exploration (PACE) framework." *Trends in Cognitive Sciences* vol. 23, 12 (2019): 1014–1025. DOI: 10.1016/j.tics.2019.10.003.

Gzielo, K., et al. "The impact of the ketogenic diet on glial cells morphology: A quantitative morphological analysis." *Neuroscience* vol. 413 (2019): 239–251. DOI: 10.1016/j.neuroscience.2019.06.009.

Hagger, Martin S., et al. "Managing stress during the coronavirus disease 2019 pandemic and beyond: Reappraisal and mindset approaches." *Stress and Health: Journal of the International Society for the Investigation of Stress* vol. 36, 3 (2020): 396–401. DOI: 10.1002/smi.2969.

Haier, Richard J. "Increased intelligence is a myth (so far)." *Frontiers in Systems Neuroscience* vol. 8, 34 (2014). DOI: 10.3389/fnsys.2014.00034.

Halgren, Milan, et al. "The generation and propagation of the human alpha rhythm." *Proceedings of the National Academy of Sciences of the United States of America* vol. 116, 47 (2019): 23772–23782. DOI: 10.1073/pnas.1913092116.

Hambrick, David, et al. (2020). "Problem-Solving and Intelligence." In *The Cambridge Handbook of Intelligence*, edited by R. Sternberg, 553–579. Cambridge Handbooks in Psychology. Cambridge, UK: Cambridge University Press. DOI: 10.1017/9781108770422 .024.

Hampson, Robert E., et al. "Developing a hippocampal neural prosthetic to facilitate human memory encoding and recall." *Journal of Neural Engineering* vol. 15, 3 (2018). DOI: 10.1088/1741-2552/aaaed7.

Harris, Alexandra, et al. "Measuring intelligence with the Sandia Matrices: Psychometric review and recommendations for free raven-like item sets." United States: N.p., 2020. DOI: doi.org/10.25035/pad.2020.03.006.

Hartman, Adam L., et al. "The neuropharmacology of the ketogenic diet." *Pediatric Neurology* vol. 36, 5 (2007): 281–292. DOI: 10.1016/j.pediatrneurol.2007.02.008.

Hassevoort, Kelsey M., et al. "Added sugar and dietary fiber consumption are associated with creativity in preadolescent children." *Nutritional Neuroscience* vol. 23, 10 (2020): 791–802. DOI: 10.1080/1028415X.2018.1558003.

Healy, C. J. "The acute effects of classic psychedelics on memory in humans." *Psychopharmacology* vol. 238, 3 (2021): 639–653. DOI: 10.1007/s00213-020-05756-w.

Heberlein, Andrea S., and Ralph Adolphs. "Impaired spontaneous anthropomorphizing despite intact perception and social knowledge." *Proceedings of the National Academy of Sciences of the United States of America* vol. 101, 19 (2004): 7487–7491. DOI: 10.1073 /pnas.0308220101.

Hegelund, Emilie Rune, et al. "The influence of educational attainment on intelligence." *Intelligence* vol. 78 (2020). http://doi.org/10.1016/j.intell.2019.101419.

Heller, Aaron S., and Rosemary C. Bagot. "Is hippocampal replay a mechanism for anxiety and depression?" *JAMA Psychiatry* vol. 77, 4 (2020): 431–432. DOI: 10.1001 /jamapsychiatry.2019.4788.

Henriksen, Danah, et al. "Mindfulness and creativity: Implications for thinking and learning." *Thinking Skills and Creativity* vol. 37 (2020). DOI: 10.1016/j.tsc.2020.100689.

Herculano-Houzel, Suzana. "Birds do have a brain cortex—and think." *Science* vol. 369, 6511 (2020): 1567–1568. DOI: 10.1126/science.abe0536.

Herculano-Houzel, Suzana. "What is so special about the human brain?" TED Talk, 2013. www.ted.com/talks/suzana_herculano_houzel_what_is_so_special_about_the _human_brain.

Higgins, Cameron, et al. "Replay bursts in humans coincide with activation of the default mode and parietal alpha networks." *Neuron* vol. 109, 5 (2021): 882–893.e7. DOI: 10.1016/j.neuron.2020.12.007.

Hilgard, Joseph, et al. "Overestimation of action-game training effects: Publication bias and salami slicing." *Collabra: Psychology* vol. 5, 1 (2019): 30. DOI: 10.1525/collabra.231.

Ho, S. Shaun, et al. "Compassion as an intervention to attune to universal suffering of self and others in conflicts: A translational framework." *Frontiers in Psychology* vol. 11 (2021). DOI: 10.3389/fpsyg.2020.603385.

Hoel, Erik. "Dream power." *New Scientist* vol. 248, 3307 (2020): 34–38.

Hoffmann, Knut, et al. "The role of dietary supplements in depression and anxiety: A narrative review." *Pharmacopsychiatry* vol. 52, 6 (2019): 261–279. DOI: 10.1055/a-0942-1875.

Hollon, Steven D. "Is cognitive therapy enduring or antidepressant medications iatrogenic? Depression as an evolved adaptation." *American Psychologist* vol. 75, 9 (2020): 1207–1218. DOI: 10.1037/amp0000728.

Holt-Lunstad, Julianne, et al. "Loneliness and social isolation as risk factors for mortality: A meta-analytic review." *Perspectives on Psychological Science: A Journal of the Association for Psychological Science* vol. 10, 2 (2015): 227–237. DOI: 10.1177/1745691614568352.

Holt-Lunstad, Julianne, et al. "Social relationships and mortality risk: A meta-analytic review." *PLoS medicine* vol. 7, 7 (2010). DOI: 10.1371/journal.pmed.1000316.

Horne, Kristina S., et al. "Evidence against benefits from cognitive training and transcranial direct current stimulation in healthy older adults." *Nature Human Behaviour* vol. 5, 1 (2021): 146–158. DOI: 10.1038/s41562-020-00979-5.

Horschig, Jörn M., et al. "Hypothesis-driven methods to augment human cognition by optimizing cortical oscillations." *Frontiers in Systems Neuroscience* vol. 8, 119 (2014). DOI: 10.3389/fnsys.2014.00119.

"How big is a billion?" University of California, Berkeley. Accessed May 10, 2021. https://ucmp.berkeley.edu/education/explorations/tours/geotime/guide/billion.html.

Huang, Hong, et al. "Physical exercise increases peripheral brain-derived neurotrophic factors in patients with cognitive impairment: A meta-analysis." *Restorative Neurology and Neuroscience* (2021). DOI: 10.3233/RNN-201060.

Hundal, Harneet, et al. "The effects of cannabidiol on persecutory ideation and anxiety in a high trait paranoid group." *Journal of Psychopharmacology* vol. 32, 3 (2018): 276–282. DOI: 10.1177/0269881117737400.

Hutten, Nadia R. P. W., et al. "Mood and cognition after administration of low LSD doses in healthy volunteers: A placebo controlled dose-effect finding study." *European Neuropsychopharmacology: The Journal of the European College of Neuropsychopharmacology* vol. 41 (2020): 81–91. DOI: 10.1016/j.euroneuro.2020.10.002.

Ide, Gladys G. "The increase of the intelligence quotient through training." *Psychological Clinic* vol. 14, 5–6 (1922): 159–162.

İnce, Rümeysa, et al. "The inventor of electroencephalography (EEG): Hans Berger (1873–1941)." *Child's Nervous System: Official Journal of the International Society for Pediatric Neurosurgery* (2020). DOI: 10.1007/s00381-020-04564-z.

Jacob, Yael, et al. "Neural correlates of rumination in major depressive disorder: A brain network analysis." *NeuroImage: Clinical* vol. 25 (2020). DOI: 10.1016/j.nicl.2019.102142.

Jacobson, A. C. "Tuberculosis and the creative mind." *Aesculapian* vol. 1, 1 (1908): 22–33.

Jasanoff, Alan. *The Biological Mind: How Brain, Body, and Environment Collaborate to Make Us Who We Are.* New York: Basic Books, 2018.

Jaschik, Scott. "The Cattell Controversy." *Inside Higher Ed*, March 20, 2009. www.insidehighered.com/news/2009/03/20/cattell-controversy.

Jenkins, Adrianna C. "Rethinking cognitive load: A default-mode network perspective." *Trends in Cognitive Sciences* vol. 23, 7 (2019): 531–533. DOI: 10.1016/j.tics.2019.04.008.

Jiang, Linxing, et al. "BrainNet: A multi-person brain-to-brain interface for direct collaboration between brains." *Scientific Reports* vol. 9, 1 (2019). DOI: 10.1038/s41598-019-41895-7.

Jiang, Yang, et al. "Tuning up the old brain with new tricks: Attention training via neurofeedback." *Frontiers in Aging Neuroscience* vol. 9, 52 (2017). DOI: 10.3389/fnagi.2017.00052.

Johnson, Elizabeth L., et al. "Insights into human cognition from intracranial EEG: A review of audition, memory, internal cognition, and causality." *Journal of Neural Engineering* vol. 17, 5 (2020). DOI: 10.1088/1741-2552/abb7a5.

Jones, David T., et al. "Default mode network disruption secondary to a lesion in the anterior thalamus." *Archives of Neurology* vol. 68, 2 (2011): 242–247. DOI: 10.1001/archneurol.2010.259.

Joshi, Sonalee A., et al. "A review of hippocampal activation in post-traumatic stress disorder." *Psychophysiology* vol. 57, 1 (2020). DOI: 10.1111/psyp.13357.

Jwa, Anita. "DIY tDCS: A need for an empirical look." *Journal of Responsible Innovation* vol. 5, 1 (2018): 103–108. DOI: 10.1080/23299460.2017.1338103.

Kadakia, Kushal T., et al. "Leveraging open science to accelerate research." *New England Journal of Medicine* (2021). DOI: 10.1056/NEJMp2034518.

Kadriu, Bashkim, et al. "Ketamine and serotonergic psychedelics: Common mechanisms underlying the effects of rapid-acting antidepressants." *International Journal of Neuropsychopharmacology* vol. 24, 1 (2021): 8–21. DOI: 10.1093/ijnp/pyaa087.

Kamp, M. A., et al. "Traumatic brain injuries in the ancient Egypt: Insights from the Edwin Smith Papyrus." *Journal of Neurological Surgery: Part A, Central European Neurosurgery* vol. 73, 4 (2012): 230–237. DOI: 10.1055/s-0032-1313635.

Kan, Kees-Jan, et al. "Extending psychometric network analysis: Empirical evidence against *g* in favor of mutualism?" *Intelligence* vol. 73 (2019): 52–62. DOI: 10.1016/j.intell.2018.12.004.

Kandola, Aaron, et al. "Physical activity and depression: Towards understanding the antidepressant mechanisms of physical activity." *Neuroscience and Biobehavioral Reviews* vol. 107 (2019): 525–539. DOI: 10.1016/j.neubiorev.2019.09.040.

Kaplan, Jonas T., et al. "Processing narratives concerning protected values: A cross-cultural investigation of neural correlates." *Cerebral Cortex* vol. 27, 2 (2017): 1428–1438. DOI: 10.1093/cercor/bhv325.

Katsumi, Yuta, et al. "Intrinsic functional network contributions to the relationship between trait empathy and subjective happiness." *NeuroImage* vol. 227 (2021). DOI: 10.1016/j.neuroimage.2020.117650.

Kaufman, James. "Creativity as a stepping stone toward a brighter future." *Journal of Intelligence* vol. 6, 2 (2018). DOI: 10.3390/jintelligence6020021.

Ke, Yufeng, et al. "The effects of transcranial direct current stimulation (tDCS) on working memory training in healthy young adults." *Frontiers in Human Neuroscience* vol. 13 (2019). DOI: 10.3389/fnhum.2019.00019.

Keltner, D., and G. A. Bonanno. "A study of laughter and dissociation: Distinct correlates of laughter and smiling during bereavement." *Journal of Personality and Social Psychology* vol. 73, 4 (1997): 687–702. DOI: 10.1037//0022-3514.73.4.687.

Kinreich, Sivan, et al. "Brain-to-brain synchrony during naturalistic social interactions." *Scientific Reports* vol. 7, 1 (2017). DOI: 10.1038/s41598-017-17339-5.

Krystal, John H., et al. "Ketamine: A paradigm shift for depression research and treatment." *Neuron* vol. 101, 5 (2019): 774–778. DOI: 10.1016/j.neuron.2019.02.005.

Krznaric, Roman. "Six habits of highly empathic people." *Greater Good Magazine*, November 27, 2012. https://greatergood.berkeley.edu/article/item/six_habits_of_highly _empathic_people1.

Kühn, Simone, et al. "Effects of computer gaming on cognition, brain structure, and function: A critical reflection on existing literature." *Dialogues in Clinical Neuroscience* vol. 21, 3 (2019): 319–330. DOI: 10.31887/DCNS.2019.21.3/skuehn.

Kuypers, Kim P. C., et al. "Microdosing psychedelics: More questions than answers? An overview and suggestions for future research." *Journal of Psychopharmacology* vol. 33, 9 (2019): 1039–1057. DOI: 10.1177/0269881119857204.

Laland, Kevin M. "These amazing creative animals show why humans are the most innovative species of all." *The Conversation*, April 20, 2017. https://theconversation.com /these-amazing-creative-animals-show-why-humans-are-the-most-innovative-species-of -all-75515.

Lanius, Ruth A., et al. "The sense of self in the aftermath of trauma: Lessons from the default mode network in posttraumatic stress disorder." *European Journal of Psychotraumatology* vol. 11, 1 (2020). DOI: 10.1080/20008198.2020.1807703.

Larsen, Christian, and Jorida Shahinas. "Dosage, efficacy and safety of cannabidiol administration in adults: A systematic review of human trials." *Journal of Clinical Medicine Research* vol. 12, 3 (2020): 129–141. DOI: 10.14740/jocmr4090.

Lavazza, Andrea. "Transcranial electrical stimulation for human enhancement and the risk of inequality: Prohibition or compensation?" *Bioethics* vol. 33, 1 (2018): 122–131. DOI: 10.1111/bioe.12504.

Lavretsky, Helen, and Jack L. Feldman. "Precision medicine for breath-focused mind-body therapies for stress and anxiety: Are we ready yet?" *Global Advances in Health and Medicine* vol. 10 (2021). DOI: 10.1177/2164956120986129.

Leblanc, Heloise, and Steve Ramirez. "Linking social cognition to learning and memory." *Journal of Neuroscience: The Official Journal of the Society for Neuroscience* vol. 40, 46 (2020): 8782–8798. DOI: 10.1523/JNEUROSCI.1280-20.2020.

Lebuda, Izabela, et al. "Mind full of ideas: A meta-analysis of the mindfulness–creativity link." *Personality and Individual Differences* vol. 93 (2016): 22–26.

Leptourgos, Pantelis, et al. "Hallucinations under psychedelics and in the schizophrenia spectrum: An interdisciplinary and multiscale comparison." *Schizophrenia Bulletin* vol. 46, 6 (2020): 1396–1408. DOI: 10.1093/schbul/sbaa117.

Li, Wanqing, et al. "The default mode network and social understanding of others: What do brain connectivity studies tell us?" *Frontiers in Human Neuroscience* vol. 8 (2014). DOI: 10.3389/fnhum.2014.00074.

Li, Wenjuan, et al. "Dual n-back working memory training evinces superior transfer effects compared to the method of loci." *Scientific Reports* vol. 11, 1 (2021). DOI: 10.1038 /s41598-021-82663-w.

Li, Yamei, et al. "Cerebral functional manipulation of repetitive transcranial magnetic stimulation in cognitive impairment patients after stroke: An fMRI study." *Frontiers in Neurology* vol. 11 (2020). DOI: 10.3389/fneur.2020.00977.

Liao, Yuhua, et al. "Efficacy of omega-3 PUFAs in depression: A meta-analysis." *Translational Psychiatry* vol. 9, 1 (2019). DOI: 10.1038/s41398-019-0515-5.

Liao, Yanhui, et al. "Brief mindfulness-based intervention of 'STOP (Stop, Take a Breath, Observe, Proceed) touching your face': A study protocol of a randomised controlled trial." *BMJ Open* vol. 10, 11 (2020). DOI: 10.1136/bmjopen-2020-041364.

Lisman, John E., and Ole Jensen. "The θ-γ neural code." *Neuron* vol. 77, 6 (2013): 1002–1016. DOI: 10.1016/j.neuron.2013.03.007.

Liu, Jenny J. W., et al. "The efficacy of stress reappraisal interventions on stress responsivity: A meta-analysis and systematic review of existing evidence." *PloS One* vol. 14, 2 (2019). DOI: 10.1371/journal.pone.0212854.

Liu, Shijie, et al. "Effects of acute and chronic exercises on executive function in children and adolescents: A systemic review and meta-analysis." *Frontiers in Psychology* vol. 11: 554915 (2020). DOI: 10.3389/fpsyg.2020.554915.

Liu, Xinyi, et al. "Disrupted rich-club network organization and individualized identification of patients with major depressive disorder." *Progress in Neuro-Psychopharmacology and Biological Psychiatry* vol. 108 (2021). DOI: 10.1016/j.pnpbp.2020.110074.

Lopez, Cesar Augusto Fontanillo, et al. "Beyond technologies of electroencephalography-based brain-computer interfaces: A systematic review from commercial and ethical aspects." *Frontiers in Neuroscience* vol. 14 (2020). DOI: 10.3389/fnins.2020.611130.

López-López, José A., et al. "The process and delivery of CBT for depression in adults: A systematic review and network meta-analysis." *Psychological Medicine* vol. 49, 12 (2019): 1937–1947. DOI: 10.1017/S003329171900120X.

Lovaas, O. Ivar. *Teaching Developmentally Disabled Children: The Me Book*. Austin: PRO-ED, 1981.

Luber, Bruce, and Sarah H. Lisanby. "Enhancement of human cognitive performance using transcranial magnetic stimulation (TMS)." *NeuroImage* vol. 85 (2014): 961–970. DOI: 10.1016/j.neuroimage.2013.06.007.

Luberto, Christina M., et al. "A systematic review and meta-analysis of the effects of meditation on empathy, compassion, and prosocial behaviors." *Mindfulness* vol. 9, 3 (2018): 708–724. DOI: 10.1007/s12671-017-0841-8.

Ma, Hsen-Hsing. "The effect size of variables associated with creativity: A meta-analysis." *Creativity Research Journal* vol. 21, 1 (2009) 30–42. DOI: 10.1080/10400410802633400.

MacLean, Katherine A., et al. "Mystical experiences occasioned by the hallucinogen psilocybin lead to increases in the personality domain of openness." *Journal of Psychopharmacology* vol. 25, 11 (2011): 1453–1461. DOI: 10.1177/0269881111420188.

Madore, Kevin P., et al. "Neural mechanisms of episodic retrieval support divergent creative thinking." *Cerebral Cortex* vol. 29, 1 (2019): 150–166. DOI: 10.1093/cercor/bhx312.

Malik, A. S., and H. Amin. "Designing an EEG experiment." In *Designing EEG Experiments for Studying the Brain*, by Malik and Amin. Cambridge, MA: Academic Press, 2018. DOI: 10.1016/B978-0-12-811140-6.00001-1.

Mani, Anandi, et al. "Poverty impedes cognitive function." *Science* vol. 341, 6149 (2013): 976–980. DOI: 10.1126/science.1238041.

Mankin, Emily A., and Itzhak Fried. "Modulation of human memory by deep brain stimulation of the entorhinal-hippocampal circuitry." *Neuron* vol. 106, 2 (2020): 218–235. DOI: 10.1016/j.neuron.2020.02.024.

Mar, R. A. "Stories and the promotion of social cognition." *Current Directions in Psychological Science* vol. 27, 4 (2018): 257–262. DOI: 10.1177/0963721417749654.

Mason, N. L., et al. "Me, myself, bye: Regional alterations in glutamate and the experience of ego dissolution with psilocybin." *Neuropsychopharmacology: Official Publication of the American College of Neuropsychopharmacology* vol. 45, 12 (2020): 2003–2011. DOI: 10.1038/s41386-020-0718-8.

Mason, N. L., et al. "Spontaneous and deliberate creative cognition during and after psilocybin exposure." *Translational Psychiatry* vol. 11, 1 (2021). DOI: 10.1038/s41398 -021-01335-5.

"Mastering your memory Dominic O'Brien." YouTube video. April 15, 2019. www.youtube .com/watch?v=75nJI7DGRlI.

Mayor, Eric, et al. "The Dark Triad of personality and attitudes toward cognitive enhancement." *BMC Psychology* vol. 8, 1 (2020). DOI: 10.1186/s40359-020-00486-2.

Mazurek, Kevin A., and Marc H Schieber. "Injecting information into the mammalian cortex: Progress, challenges, and promise." *The Neuroscientist: A Review Journal Bringing Neurobiology, Neurology and Psychiatry* vol. 27, 2 (2021): 129–142. DOI: 10.1177/1073858420936253.

McNamara, Robert K., and Daniel M. Almeida. "Omega-3 polyunsaturated fatty acid deficiency and progressive neuropathology in psychiatric disorders: A review of translational evidence and candidate mechanisms." *Harvard Review of Psychiatry* vol. 27, 2 (2019): 94–107. DOI: 10.1097/HRP.0000000000000199.

McPhilemy, Genevieve, et al. "A 52-week prophylactic randomised control trial of omega-3 polyunsaturated fatty acids in bipolar disorder." *Bipolar Disorders* (2020). DOI: 10.1111 /bdi.13037.

Melby-Lervåg, Monica, and Charles Hulme. "Is working memory training effective? A meta-analytic review." *Developmental Psychology* vol. 49, 2 (2013): 270–291. DOI: 10.1037 /a0028228.

Mende, Melinda A., and Hendrikje Schmidt. "Psychotherapy in the framework of embodied cognition: Does interpersonal synchrony influence therapy success?" *Frontiers in Psychiatry* vol. 12 (2021). DOI: 10.3389/fpsyt.2021.562490.

Meredith, Genevive R., et al. "Minimum time dose in nature to positively impact the mental health of college-aged students, and how to measure it: A scoping review." *Frontiers in Psychology* vol. 10 (2020). DOI: 10.3389/fpsyg.2019.02942.

Miles, Andrew, et al. "Using prosocial behavior to safeguard mental health and foster emotional well-being during the COVID-19 pandemic: A registered report protocol for a randomized trial." *PloS One* vol. 16, 1 (2021). DOI: 10.1371/journal.pone .0245865.

Minielly, Nicole, et al. "Privacy challenges to the democratization of brain data." *iScience* vol. 23, 6 (2020). DOI: 10.1016/j.isci.2020.101134.

Mitchell, Jennifer M., et al. "MDMA-assisted therapy for severe PTSD: A randomized, double-blind, placebo-controlled phase 3 study." *Nature Medicine* (2021). DOI: 10.1038 /s41591-021-01336-3.

Momi, D., et al. "Cognitive enhancement via network-targeted cortico-cortical associative brain stimulation." *Cerebral Cortex* vol. 30, 3 (2020): 1516–1527. DOI: 10.1093/cercor /bhz182.

Monson, Candice M., et al. "MDMA-facilitated cognitive-behavioural conjoint therapy for posttraumatic stress disorder: An uncontrolled trial." *European*

Journal of Psychotraumatology vol. 11, 1 (2020). DOI: 10.1080/20008198.2020.184
0123.

Moore, Malcolm. "Huge dam may have triggered Sichuan earthquake, scientists say." *Sydney Morning Herald*, February 4, 2009. www.smh.com.au/world/huge-dam-may-have
-triggered-sichuan-earthquake-scientists-say-20090204-gdtc8u.html.

Moskowitz, Judith T., et al. "Randomized controlled trial of a positive affect intervention for people newly diagnosed with HIV." *Journal of Consulting and Clinical Psychology* vol. 85, 5 (2017): 409–423. DOI: 10.1037/ccp0000188.

Munjuluri, Sarat, et al. "A pilot study on playback theatre as a therapeutic aid after natural disasters: Brain connectivity mechanisms of effects on anxiety." *Chronic Stress* vol. 4 (2020). DOI: 10.1177/2470547020966561.

Murphy, Andrew C., et al. "Multimodal network dynamics underpinning working memory." *Nature Communications* vol. 11, 1 (2020). DOI: 10.1038/s41467-020-15541-0.

Musk, Elon, and Neuralink. "An integrated brain-machine interface platform with thousands of channels." *Journal of Medical Internet Research* vol. 21,10 (2019). DOI: 10.2196/16194.

Nęcka, Edward, et al. "The effects of working memory training on brain activity." *Brain Sciences* vol. 11, 2 (2021). DOI: 10.3390/brainsci11020155.

Neggers, Sebastiaan F. W., et al. "Understanding the biophysical effects of transcranial magnetic stimulation on brain tissue: The bridge between brain stimulation and cognition." *Progress in Brain Research* vol. 222 (2015): 229–259. DOI: 10.1016/bs.pbr.2015.06.015.

Nelson, S. Katherine, et al. "Do unto others or treat yourself? The effects of prosocial and self-focused behavior on psychological flourishing." *Emotion* vol. 16, 6 (2016): 850–861. DOI: 10.1037/emo0000178.

Neubauer, Simon, et al. "The evolution of modern human brain shape." *Science Advances* vol. 4, 1 (2018). DOI: 10.1126/sciadv.aao5961.

Nicholson, Andrew A., et al. "A randomized, controlled trial of alpha-rhythm EEG neurofeedback in posttraumatic stress disorder: A preliminary investigation showing evidence of decreased PTSD symptoms and restored default mode and salience network connectivity using fMRI." *NeuroImage: Clinical* vol. 28 (2020). DOI: 10.1016/j.nicl.2020.102490.

Nicolelis, Miguel. *The True Creator of Everything: How the Human Brain Shaped the Universe as We Know It*. New Haven, CT: Yale University Press, 2020.

Oberste, Max, et al. "Acute exercise-induced set shifting benefits in healthy adults and its moderators: A systematic review and meta-analysis." *Frontiers in Psychology* vol. 12 (2021). DOI: 10.3389/fpsyg.2021.528352.

Okereke, Olivia I., et al. "Effect of long-term vitamin D3 supplementation vs placebo on risk of depression or clinically relevant depressive symptoms and on change in mood scores: A randomized clinical trial." *JAMA* vol. 324, 5 (2020): 471–480. DOI: 10.1001
/jama.2020.10224.

O'Leary, Karen, and Samantha Dockray. "The effects of two novel gratitude and mindfulness interventions on well-being." *Journal of Alternative and Complementary Medicine* vol. 21, 4 (2015): 243–245. DOI: 10.1089/acm.2014.0119.

Oró, Pere, et al. "Effectiveness of a mindfulness-based programme on perceived stress, psychopathological symptomatology and burnout in medical students." *Mindfulness* vol. 12 (2021). DOI: 10.1007/s12671-020-01582-5.

Orsolini, Laura, et al. "Psychedelic fauna for psychonaut hunters: A mini-review." *Frontiers in Psychiatry* vol. 9 (2018). DOI: 10.3389/fpsyt.2018.00153.

Oxley, Thomas J., et al. "Motor neuroprosthesis implanted with neurointerventional surgery improves capacity for activities of daily living tasks in severe paralysis: First in-human experience." *Journal of Neurointerventional Surgery* vol. 13, 2 (2021): 102–108. DOI: 10.1136/neurintsurg-2020-016862.

Parkinson, Carolyn, et al. "Similar neural responses predict friendship." *Nature Communications* vol. 9, 1 (2018). DOI: 10.1038/s41467-017-02722-7.

Pattee, Emma. "5 people who can help you strengthen your empathy muscle." *New York Times*, October 4, 2020. www.nytimes.com/2020/10/04/smarter-living/5-people-who -can-help-you-strengthen-your-empathy-muscle.html.

Pavón, S., et al. "Ketogenic diet and cognition in neurological diseases: A systematic review." *Nutrition Reviews* (2020). DOI: 10.1093/nutrit/nuaa113.

Peetz, Johanna, and Igor Grossmann. "Wise reasoning about the future is associated with adaptive interpersonal feelings after relational challenges." *Social Psychological and Personality Science* (2020). DOI: 10.1177/1948550620931985.

Pennycook, Gordon, et al. "On the belief that beliefs should change according to evidence: Implications for conspiratorial, moral, paranormal, political, religious, and science beliefs." *Judgment and Decision Making* vol. 15, 4 (2020): 476–498.

Perchtold-Stefan, Corinna M., et al. "More habitual physical activity is linked to the use of specific, more adaptive cognitive reappraisal strategies in dealing with stressful events." *Stress and Health: Journal of the International Society for the Investigation of Stress* vol. 36, 3 (2020): 274–286. DOI: 10.1002/smi.2929.

Perou, Ruth, et al. "The Legacy for Children™ randomized control trial: Effects on cognition through third grade for young children experiencing poverty." *Journal of Developmental and Behavioral Pediatrics* vol. 40, 4 (2019): 275–284. DOI: 10.1097 /DBP.0000000000000656.

Perri, Rinaldo Livio, et al. "COVID-19, isolation, quarantine: On the efficacy of internet-based eye movement desensitization and reprocessing (EMDR) and Cognitive-Behavioral Therapy (CBT) for ongoing trauma." *Brain Sciences* vol. 11, 5 (2021). DOI: 10.3390/ brainsci11050579.

Petrosino, Nicholas J., et al. "One-year clinical outcomes following theta burst stimulation for post-traumatic stress disorder." *Neuropsychopharmacology: Official Publication of the American College of Neuropsychopharmacology* vol. 45, 6 (2020): 940–946. DOI: 10.1038 /s41386-019-0584-4.

Peveretou, Foteini, et al. "A short empathy paradigm to assess empathic deficits in schizophrenia." *Behavioral Sciences* vol. 10, 2 (2020). DOI: 10.3390/bs10020041.

Plailly, J., et al. "Incorporation of fragmented visuo-olfactory episodic memory into dreams and its association with memory performance." *Scientific Reports* vol. 9, 1 (2019). DOI: 10.1038/s41598-019-51497-y.

Porter, Blake. "*Inside Out*'s take on the brain: A neuroscientist's perspective." *Dr. Blake Porter* (blog). June 27, 2015. www.blakeporterneuro.com/inside-outs-take-on-the-brain -a-neuroscientists-perspective/.

Preiss, David D., and Diego Cosmelli. "Mind wandering, creative writing, and the self." In *Explorations in Creativity Research: The Creative Self: Effect of Beliefs, Self-Efficacy,*

Mindset, and Identity, edited by M. Karwowski and J. C. Kaufman, 301–313. Cambridge, MA: Elsevier Academic Press, 2017. DOI: 10.1016/B978-0-12-809790-8.00017-0.

Preller, Katrin H., and Franz X. Vollenweider. "Modulation of social cognition via hallucinogens and 'entactogens.'" *Frontiers in Psychiatry* vol. 10 (2019). DOI: 10.3389/fpsyt.2019.00881.

Pugh, Jonathan, et al. "Brainjacking in deep brain stimulation and autonomy." *Ethics and Information Technology* vol. 20, 3 (2018): 219–232. DOI: 10.1007/s10676-018-9466-4.

Purslow, Matt. "Elon Musk's brain company hopes to make a cyborg monkey play 'Mind Pong.'" IGN. February 1, 2021. www.ign.com/articles/elon-musk-brain-implant-monkey-neuralink.

Qiao, Lei, et al. "Spontaneous brain state oscillation is associated with self-reported anxiety in a non-clinical sample." *Scientific Reports* vol. 10, 1 (2020). DOI: 10.1038/s41598-020-76211-1.

Raichle, Marcus E. "The brain's default mode network." *Annual Review of Neuroscience* vol. 38, 1 (2015): 433–447.

Ramirez-Zamora, Adolfo, et al. "Proceedings of the seventh annual Deep Brain Stimulation Think Tank: Advances in Neurophysiology, Adaptive DBS, Virtual Reality, Neuroethics and Technology." *Frontiers in Human Neuroscience* vol. 14 (2020). DOI: 10.3389/fnhum.2020.00054.

Raposo Pereira, Filipa, et al. "Recreational use of GHB is associated with alterations of resting state functional connectivity of the central executive and default mode networks." *Human Brain Mapping* vol. 40, 8 (2019): 2413–2421. DOI: 10.1002/hbm.24532.

Ravindran, Arun V., et al. "Canadian Network for Mood and Anxiety Treatments (CANMAT) 2016 clinical guidelines for the management of adults with major depressive disorder: Section 5, complementary and alternative medicine treatments." *Canadian Journal of Psychiatry: Revue canadienne de psychiatrie* vol. 61, 9 (2016): 576–587. DOI: 10.1177/0706743716660290.

Rawat, Kajal, et al. "A review on preventive role of ketogenic diet (KD) in CNS disorders from the gut microbiota perspective." *Reviews in the Neurosciences* vol. 32, 2 (2020): 143–157. DOI: 10.1515/revneuro-2020-0078.

Raymond, Catherine, et al. "Should we suppress or reappraise our stress? The moderating role of reappraisal on cortisol reactivity and recovery in healthy adults." *Anxiety, Stress, and Coping* vol. 32, 3 (2019): 286–297. DOI: 10.1080/10615806.2019.1596676.

Reinhart, Robert M. G., et al. "Using transcranial direct-current stimulation (tDCS) to understand cognitive processing." *Attention, Perception and Psychophysics* vol. 79, 1 (2017): 3–23. DOI: 10.3758/s13414-016-1224-2.

Repantis, Dimitris, et al. "Cognitive enhancement effects of stimulants: A randomized controlled trial testing methylphenidate, modafinil, and caffeine." *Psychopharmacology* vol. 238, 2 (2021): 441–451. DOI: 10.1007/s00213-020-05691-w.

Repantis, Dimitris, et al. "Modafinil and methylphenidate for neuroenhancement in healthy individuals: A systematic review." *Pharmacological Research* vol. 62, 3 (2010): 187–206. DOI: 10.1016/j.phrs.2010.04.002.

Restrepo-Martínez, Miguel, et al. "FDG-PET in Cotard syndrome before and after treatment: Can functional brain imaging support a two-factor hypothesis of nihilistic delusions?" *Cognitive Neuropsychiatry* vol. 24, 6 (2019): 470–480. DOI: 10.1080/13546805.2019.1676710.

Ricci, Alessandro, et al. "Exploring the mechanisms of action of the antidepressant effect of the ketogenic diet." *Reviews in the Neurosciences* vol. 31, 6 (2020): 637–648. DOI: 10.1515/revneuro-2019-0073.

Richardson, Ken, and Sarah H. Norgate. "Does IQ measure ability for complex cognition?" *Theory and Psychology* vol. 24, 6 (2014): 795–812. DOI: 10.1177/0959354314551163.

Risen, Clay. "James R. Flynn, who found we are getting smarter, dies at 86." *New York Times*, January 25, 2021. www.nytimes.com/2021/01/25/science/james-r-flynn-dead.html.

Ritchie, Stuart J., and Elliot M. Tucker-Drob. "How much does education improve intelligence? A meta-analysis." *Psychological Science* vol. 29, 8 (2018): 1358–1369. DOI: 10.1177/0956797618774253.

Robillard, Julie M., et al. "Scientific and ethical features of English-language online tests for Alzheimer's disease." *Alzheimer's and Dementia* vol. 1, 3 (2015): 281–288. DOI: 10.1016/j.dadm.2015.03.004.

Rogers, Adam. "Neuralink is impressive tech, wrapped in Musk hype." *Wired*, September 4, 2020. www.wired.com/story/neuralink-is-impressive-tech-wrapped-in-musk-hype/.

Roscoe, Zoe. "How many Super Bowls has Tom Brady won?" *Parade*, February 7, 2021. https://parade.com/976983/zoeroscoe/how-many-super-bowls-has-tom-brady-won/.

Rosen, David S., et al. "Anodal tDCS to right dorsolateral prefrontal cortex facilitates performance for novice jazz improvisers but hinders experts." *Frontiers in Human Neuroscience* vol. 10, 579 (2016). DOI: 10.3389/fnhum.2016.00579.

Ruminjo, Anne, and Boris Mekinulov. "A case report of Cotard's syndrome." *Psychiatry* vol. 5, 6 (2008): 28–29.

Sala, Giovanni, and Fernand Gobet. "Cognitive training does not enhance general cognition." *Trends in Cognitive Sciences* vol. 23, 1 (2019): 9–20. DOI: 10.1016/j.tics.2018.10.004.

Salehi, Mehraveh, et al. "Individualized functional networks reconfigure with cognitive state." *NeuroImage* vol. 206 (2020). DOI: 10.1016/j.neuroimage.2019.116233.

Salvi, Carola, et al. "TDCS to the right anterior temporal lobe facilitates insight problem-solving." *Scientific Reports* vol. 10, 1 (2020). DOI: 10.1038/s41598-020-57724-1.

Sanchez, Justin, and Robbin Miranda. "Taking neurotechnology into new territory." Defense Media Network. March 14, 2019. www.defensemedianetwork.com/stories/taking-neurotechnology-new-territory/.

Sani, Omid G., et al. "Mood variations decoded from multi-site intracranial human brain activity." *Nature Biotechnology* vol. 36, 10 (2018): 954–961. DOI: 10.1038/nbt.4200.

Saris, Ilja M. J., et al. "Default mode network connectivity and social dysfunction in major depressive disorder." *Scientific Reports* vol. 10, 1 (2020). DOI: 10.1038/s41598-019-57033-2.

Schmid, Yasmin, et al. "Acute effects of lysergic acid diethylamide in healthy subjects." *Biological Psychiatry* vol. 78, 8 (2015): 544–553. DOI: 10.1016/j.biopsych.2014.11.015.

Schneider, Felicitas, et al. "Delaying memory decline: Different options and emerging solutions." *Translational Psychiatry* vol. 10, 1 (2020). DOI: 10.1038/s41398-020-0697-x.

Schuman-Olivier, Zev, et al. "Mindfulness and behavior change." *Harvard Review of Psychiatry* vol. 28, 6 (2020): 371–394. DOI: 10.1097/HRP.0000000000000277.

Segal, Nancy L. "Twins, birth weight, cognition, and handedness." *Proceedings of the National Academy of Sciences of the United States of America* vol. 109, 48 (2012): E3293; author reply E3294. DOI: 10.1073/pnas.1213701109.

Selhub, Eva. "Nutritional psychiatry: Your brain on food." *Harvard Health Blog.* Updated March 26, 2020. www.health.harvard.edu/blog/nutritional-psychiatry-your-brain-on-food-201511168626.

Selita, Fatos, and Yulia Kovas. "Genes and Gini: What inequality means for heritability." *Journal of Biosocial Science* vol. 51, 1 (2019): 18–47. DOI:10.1017/S0021932017000645.

Shadick, Nancy A., et al. "A randomized controlled trial of an internal family systems-based psychotherapeutic intervention on outcomes in rheumatoid arthritis: A proof-of-concept study." *Journal of Rheumatology* vol. 40, 11 (2013): 1831–1841. DOI: 10.3899/jrheum.121465.

Shannon, Scott, et al. "Cannabidiol in anxiety and sleep: A large case series." *Permanente Journal* vol. 23 (2019). DOI: 10.7812/TPP/18-041.

Shen, Xueyi, et al. "Resting-state connectivity and its association with cognitive performance, educational attainment, and household income in the UK Biobank." *Biological Psychiatry* vol. 3, 10 (2018): 878–886. DOI: 10.1016/j.bpsc.2018.06.007.

Sholler, Dennis J., et al. "Therapeutic efficacy of cannabidiol (CBD): A review of the evidence from clinical trials and human laboratory studies." *Current Addiction Reports* vol. 7, 3 (2020): 405–412. DOI: 10.1007/s40429-020-00326-8.

Simons, Daniel J., et al. "Do "brain-training" programs work?" *Psychological Science in the Public Interest: A Journal of the American Psychological Society* vol. 17, 3 (2016): 103–186. DOI: 10.1177/1529100616661983.

Sims, Regina C., et al. "The influence of functional social support on executive functioning in middle-aged African Americans." *Neuropsychology, Development, and Cognition: Section B, Aging, Neuropsychology and Cognition* vol. 18, 4 (2011): 414–431. DOI: 10.1080/13825585.2011.567325.

Smith, Caroline A., et al. "Acupuncture for depression." *Cochrane Database of Systematic Reviews* vol. 3, 3 (2018). DOI: 10.1002/14651858.CD004046.pub4.

Spanò, Goffredina, et al. "Dreaming with hippocampal damage." *eLife* vol. 9 (2020). DOI: 10.7554/eLife.56211.

Sparling, Jessica E., et al. "Environmental enrichment and its influence on rodent offspring and maternal behaviours, a scoping style review of indices of depression and anxiety." *Pharmacology, Biochemistry, and Behavior* vol. 197 (2020). DOI: 10.1016/j.phb.2020.172997.

Spinella, Toni C., et al. "Evaluating cannabidiol (CBD) expectancy effects on acute stress and anxiety in healthy adults: A randomized crossover study." *Psychopharmacology* (2021). DOI: 10.1007/s00213-021-05823-w.

Spinelli, Alexander. "New research finds unapproved drugs in brain-boosting supplements." *STAT*, September 23, 2020. www.statnews.com/2020/09/23/unapproved-drugs-brain-boosting-supplements/.

Spunt, Robert P., et al. "The default mode of human brain function primes the intentional stance." *Journal of Cognitive Neuroscience* vol. 27, 6 (2015): 1116–1124. DOI: 10.1162/jocn_a_00785.

Spurny, B., et al. "Effects of SSRI treatment on GABA and glutamate levels in an associative relearning paradigm." *NeuroImage* vol. 232 (2021). DOI: 10.1016/j.neuroimage.2021.117913.

Sripada, Chandra, et al. "Toward a 'treadmill test' for cognition: Improved prediction of general cognitive ability from the task activated brain." *Human Brain Mapping* vol. 41, 12 (2020): 3186–3197. DOI: 10.1002/hbm.25007.

Stahl, Devan, et al. "Should DBS for psychiatric disorders be considered a form of psychosurgery? Ethical and legal considerations." *Science and Engineering Ethics* vol. 24, 4 (2018): 1119–1142. DOI: 10.1007/s11948-017-9934-y.

Stankov, Lazar. "'Kvashchev's experiment': Can we boost intelligence?" *Intelligence* vol. 10 (1986): 209–230.

Stankov, Lazar, and Jihyun Lee. "We can boost IQ: Revisiting Kvashchev's experiment." *Journal of Intelligence* vol. 8, 4 (2020). DOI: 10.3390/jintelligence8040041.

Suradom, Chawisa, et al. "Omega-3 polyunsaturated fatty acid (n-3 PUFA) supplementation for prevention and treatment of perinatal depression: A systematic review and meta-analysis of randomized-controlled trials." *Nordic Journal of Psychiatry* vol. 75, 4 (2021): 239–246. DOI: 10.1080/08039488.2020.1843710.

Szigeti, Balázs, et al. "Self-blinding citizen science to explore psychedelic microdosing." *eLife* vol. 10 (2021). DOI: 10.7554/eLife.62878.

Taiwo, Zinat, et al. "Empathy for joy recruits a broader prefrontal network than empathy for sadness and is predicted by executive functioning." *Neuropsychology* vol. 35, 1 (2021): 90–102. DOI: 10.1037/neu0000666.

Tamir, Diana I., et al. "Reading fiction and reading minds: The role of simulation in the default network." *Social Cognitive and Affective Neuroscience* vol. 11, 2 (2016): 215–224. DOI: 10.1093/scan/nsv114.

Tang, Yan, et al. "Aberrant functional brain connectome in people with antisocial personality disorder." *Scientific Reports* vol. 6 (2016). DOI: 10.1038/srep26209.

Taren, Adrienne A., et al. "Mindfulness meditation training and executive control network resting state functional connectivity: A randomized controlled trial." *Psychosomatic Medicine* vol. 79, 6 (2017): 674–683. DOI: 10.1097/PSY.0000000000000466.

Taylor, Bridget, and David Celiberti. "A tribute to Dr. Ivar Lovaas." *Science in Autism Treatment*, Fall 2010. https://asatonline.org/wp-content/uploads/NewsletterIssues/fall2010.pdf.

Taylor, Charles T., et al. "Enhancing social connectedness in anxiety and depression through amplification of positivity: Preliminary treatment outcomes and process of change." *Cognitive Therapy and Research* vol. 44, 4 (2020): 788–800. DOI: 10.1007/s10608-020-10102-7.

Thesing, Carisha S., F. Lamers, et al. "Response to 'International Society for Nutritional Psychiatry research practice guidelines for omega-3 fatty acids in the treatment of major depressive disorder' by Guu et al. (2020)." *Psychotherapy and Psychosomatics* vol. 89, 1 (2020): 48. DOI: 10.1159/000504100.

Thesing, Carisha S., Y. Milaneschi, et al. "Supplementation-induced increase in circulating omega-3 serum levels is not associated with a reduction in depressive symptoms: Results from the MooDFOOD depression prevention trial." *Depression and Anxiety* vol. 37, 11 (2020): 1079–1088. DOI: 10.1002/da.23092.

Tobin, Matthew K., et al. "Human hippocampal neurogenesis persists in aged adults and Alzheimer's disease patients." *Cell: Stem Cell* vol. 24, 6 (2019): 974–982.e3. DOI: 10.1016/j.stem.2019.05.003.

Trambaiolli, Lucas R., et al. "Neurofeedback training in major depressive disorder: A systematic review of clinical efficacy, study quality and reporting practices." *Neuroscience and Biobehavioral Reviews* vol. 125 (2021): 33–56. DOI: 10.1016/j.neubiorev.2021.02.015.

Trautwein, Fynn-Mathis, et al. "Differential benefits of mental training types for attention, compassion, and theory of mind." *Cognition* vol. 194 (2020). DOI: 10.1016/j.cognition.2019.104039.

Trevathan, James K., et al. "An injectable neural stimulation electrode made from an in-body curing polymer/metal composite." *Advanced Healthcare Materials* vol. 8, 23 (2019). DOI: 10.1002/adhm.201900892.

Tucker, William H. "The racist past of the American psychology establishment." *Journal of Blacks in Higher Education*, 2005. www.jbhe.com/features/48_Cattell_case.html.

Uddin, Lucina Q. "Cognitive and behavioural flexibility: Neural mechanisms and clinical considerations." *Nature Reviews: Neuroscience* vol. 22, 3 (2021): 167–179. DOI: 10.1038/s41583-021-00428-w.

Umbach, Rebecca H., and Nim Tottenham. "Callous-unemotional traits and reduced default mode network connectivity within a community sample of children." *Development and Psychopathology* (2020): 1–14. DOI: 10.1017/S0954579420000401.

Ungar, Michael. "Designing resilience research: Using multiple methods to investigate risk exposure, promotive and protective processes, and contextually relevant outcomes for children and youth." *Child Abuse and Neglect* vol. 96 (2019). DOI: 10.1016/j.chiabu.2019.104098.

Vallat, Raphael, et al. "Brain functional connectivity upon awakening from sleep predicts interindividual differences in dream recall frequency." *Sleep* vol. 43, 12 (2020). DOI: 10.1093/sleep/zsaa116.

Vallat, Raphael, and Perrine Marie Ruby. "Is it a good idea to cultivate lucid dreaming?" *Frontiers in Psychology* vol. 10 (2019). DOI: 10.3389/fpsyg.2019.02585.

Van den Heuvel, Martijn P., et al. "Evolutionary modifications in human brain connectivity associated with schizophrenia." *Brain: A Journal of Neurology* vol. 142, 12 (2019): 3991–4002. DOI: 10.1093/brain/awz330.

Van Elk, Michiel, et al. "The neural correlates of the awe experience: Reduced default mode network activity during feelings of awe." *Human Brain Mapping* vol. 40, 12 (2019): 3561–3574. DOI: 10.1002/hbm.24616.

Van Ettinger-Veenstra, Helene, et al. "Chronic widespread pain patients show disrupted cortical connectivity in default mode and salience networks, modulated by pain sensitivity." *Journal of Pain Research* vol. 12 (2019): 1743–1755. DOI: 10.2147/JPR.S189443.

Vann Jones, Simon Andrew, and Allison O'Kelly. "Psychedelics as a treatment for Alzheimer's disease dementia." *Frontiers in Synaptic Neuroscience* vol. 12, 34 (2020). DOI: 10.3389/fnsyn.2020.00034.

Varley, Thomas F., et al. "Differential effects of propofol and ketamine on critical brain dynamics." *PLoS Computational Biology* vol. 16, 12 (2020). DOI: 10.1371/journal.pcbi.1008418.

Vatansever, Deniz, et al. "Default mode contributions to automated information processing." *Proceedings of the National Academy of Sciences of the United States of America* vol. 114, 48 (2017): 12821–12826. DOI: 10.1073/pnas.1710521114.

Vicentini, Jéssica Elias, et al. "Subacute functional connectivity correlates with cognitive recovery six months after stroke." *NeuroImage: Clinical* vol. 29 (2021). DOI: 10.1016/j .nicl.2020.102538.

Vollenweider, Franz X., and Katrin H. Preller. "Psychedelic drugs: Neurobiology and potential for treatment of psychiatric disorders." *Nature Reviews: Neuroscience* vol. 21, 11 (2020): 611–624. DOI: 10.1038/s41583-020-0367-2.

Vrselja, Zvonimir, et al. "Restoration of brain circulation and cellular functions hours post-mortem." *Nature* vol. 568, 7752 (2019): 336–343. DOI: 10.1038/s41586-019 -1099-1.

Wagner, I. C., et al. "Durable memories and efficient neural coding through mnemonic training using the method of loci." *Science Advances* vol. 7, 10 (2021). DOI: 10.1126 /sciadv.abc7606.

Wagner, Katy, et al. "Would you be willing to zap your child's brain? Public perspectives on parental responsibilities and the ethics of enhancing children with transcranial direct current stimulation." *AJOB Empirical Bioethics* vol. 9, 1 (2018): 29–38. DOI: 10.1080/23294515.2018.1424268.

Watts, Rosalind, et al. "Patients' accounts of increased 'connectedness' and 'acceptance' after psilocybin for treatment-resistant depression." *Journal of Humanistic Psychology* vol. 57, 5 (2017): 520–564.

Waytz, Adam, and Kurt Gray. "Does online technology make us more or less sociable? A preliminary review and call for research." *Perspectives on Psychological Science: A Journal of the Association for Psychological Science* vol. 13, 4 (2018): 473–491. DOI: 10.1177/1745691617746509.

Wertheim, Julia, et al. "Enhancing spatial reasoning by anodal transcranial direct current stimulation over the right posterior parietal cortex." *Experimental Brain Research* vol. 238, 1 (2020): 181–192. DOI: 10.1007/s00221-019-05699-5.

Williams, Alex. "Can you poison your way to good health?" *New York Times*, January 1, 2021. www.nytimes.com/2021/01/01/style/self-care/kambo-tree-frog-detox.html.

Włodarczyk, Adam, et al. "Ketogenic diet for depression: A potential dietary regimen to maintain euthymia?" *Progress in Neuro-psychopharmacology and Biological Psychiatry* vol. 109 (2021). DOI: 10.1016/j.pnpbp.2021.110257.

Włodarczyk, Adam, et al. "Ketogenic diet: A dietary modification as an anxiolytic approach?" *Nutrients* vol. 12, 12 (2020). DOI: 10.3390/nu12123822.

Wolfson, Philip E., et al. "MDMA-assisted psychotherapy for treatment of anxiety and other psychological distress related to life-threatening illnesses: A randomized pilot study." *Scientific Reports* vol. 10, 1 (2020). DOI: 10.1038/s41598-020 -75706-1.

Wong, Edward. "Year after China quake, new births, old wounds." *New York Times*, May 5, 2009. www.nytimes.com/2009/05/06/world/asia/06quake.html.

Woods, Brittany K., et al. "Isolating the effects of mindfulness training across anxiety disorder diagnoses in the unified protocol." *Behavior Therapy* vol. 51, 6 (2020): 972–983. DOI: 10.1016/j.beth.2020.01.001.

Wright, Madison, et al. "Use of cannabidiol for the treatment of anxiety: A short synthesis of pre-clinical and clinical evidence." *Cannabis and Cannabinoid Research* vol. 5, 3 (2020): 191–196. DOI: 10.1089/can.2019.0052.

Wu, Xingqi, et al. "Improved cognitive promotion through accelerated magnetic stimulation." *eNeuro* vol. 8, 1 (2021). DOI: 10.1523/ENEURO.0392-20.2020.

Xie, Hua, et al. "Spontaneous and deliberate modes of creativity: Multitask eigenconnectivity analysis captures latent cognitive modes during creative thinking." Preprint posted to bioRxiv, January 3, 2021. www.biorxiv.org/content/10.1101/2020.12.31.425008v1.

Xiong, Jian, et al. "Effects of physical exercise on executive function in cognitively healthy older adults: A systematic review and meta-analysis of randomized controlled trials: Physical exercise for executive function." *International Journal of Nursing Studies* vol. 114 (2021). DOI: 10.1016/j.ijnurstu.2020.103810.

Yang, Yiyuan, et al. "Wireless multilateral devices for optogenetic studies of individual and social behaviors." *Nature Neuroscience* (2021). DOI: 10.1038/s41593-021-00849-x.

Ybarra, Oscar, et al. "Mental exercising through simple socializing: Social interaction promotes general cognitive functioning." *Personality and Social Psychology Bulletin* vol. 34, 2 (2008): 248–259. DOI: 10.1177/0146167207310454.

Yeh, Wen-Hsiu, et al. "Neurofeedback of alpha activity on memory in healthy participants: A systematic review and meta-analysis." *Frontiers in Human Neuroscience* vol. 14 (2021). DOI: 10.3389/fnhum.2020.562360.

Yeshurun, Yaara, et al. "The default mode network: Where the idiosyncratic self meets the shared social world." *Nature Reviews: Neuroscience* vol. 22, 3 (2021): 181–192. DOI: 10.1038/s41583-020-00420-w.

Yon, Daniel, et al. "Beliefs and desires in the predictive brain." *Nature Communications* vol. 11, 1 (2020). DOI: 10.1038/s41467-020-18332-9.

Young, Alexander I., et al. "Relatedness disequilibrium regression estimates heritability without environmental bias." *Nature Genetics* vol. 50, 9 (2018): 1304–1310. DOI: 10.1038/s41588-018-0178-9.

Young, Lauren M., et al. "A systematic review and meta-analysis of B vitamin supplementation on depressive symptoms, anxiety, and stress: Effects on healthy and 'at-risk' individuals." *Nutrients* vol. 11, 9 (2019). DOI: 10.3390/nu11092232.

Youngner, Stuart, and Insoo Hyun. "Pig experiment challenges assumptions around brain damage in people." *Nature* vol. 568, 7752 (2019): 302–304. DOI: 10.1038/d41586-019-01169-8.

Zahodne, Laura B., et al. "Social relations and age-related change in memory." *Psychology and Aging* vol. 34, 6 (2019): 751–765. DOI: 10.1037/pag0000369.

Zeidman, Lawrence A., et al. "New revelations about Hans Berger, father of the electroencephalogram (EEG), and his ties to the Third Reich." *Journal of Child Neurology* vol. 29, 7 (2014): 1002–1010. DOI: 10.1177/0883073813486558.

Zehr, E. Paul. "The potential transformation of our species by neural enhancement." *Journal of Motor Behavior* vol. 47, 1 (2015): 73–78. DOI: 10.1080/00222895.2014.916652.

Zhang, Li, et al. "Omega-3 fatty acids for the treatment of depressive disorders in children and adolescents: A meta-analysis of randomized placebo-controlled trials."

Child and Adolescent Psychiatry and Mental Health vol. 13, 36 (2019). DOI: 10.1186 /s13034-019-0296-x.

Zhang, Xiao-Dong, et al. "Altered default mode network configuration in posttraumatic stress disorder after earthquake: A resting-stage functional magnetic resonance imaging study." *Medicine* vol. 96, 37 (2017). DOI: 10.1097/MD.0000000000007826.

Zhou, Yongjie, et al. "Association between social participation and cognitive function among middle- and old-aged Chinese: A fixed-effects analysis." *Journal of Global Health* vol. 10, 2 (2020). DOI: 10.7189/jogh.10.020801.

Zuk, Peter, et al. "Neuroethics of neuromodulation: An update." *Current Opinion in Biomedical Engineering* vol. 8 (2018): 45–50. DOI: 10.1016/j.cobme.2018.10.003.

INDEX

Emily Willingham is a writer and biologist. She earned a BA in English before completing a PhD in biology, both at the University of Texas at Austin. After completing a postdoctoral fellowship at the University of California, San Francisco, she took a ride on the tenure track in biology before turning her attention to journalism full time, focusing on health, science, and medicine. She is a 2021–2022 Knight Science Journalism project fellow at MIT. She has served on the board of the National Association of Science Writers, and her bylines have appeared in the *New York Times*, *Scientific American*, *Forbes*, *Washington Post*, *Wall Street Journal*, *San Francisco Chronicle*, *Slate*, *Men's Health*, and elsewhere. Willingham is the coauthor of *The Informed Parent* and the author of *Phallacy*, which was recently selected for the Quarantine Book Club. She lives in California.